Brooklands Books

MOTO GUZZI

Gold Portfolio

1949 - 1973

Compiled by
R.M.Clarke

ISBN 1 85520 3790

Brooklands Books

BROOKLANDS BOOKS LTD.
P.O. BOX 146, COBHAM,
SURREY, KT11 1LG. UK

A -MI49BGP

Printed in Hong Kong

Brooklands Books

MOTORING

BROOKLANDS ROAD TEST SERIES

Abarth Gold Portfolio 1950-1971
AC Ace & Aceca 1953-1983
Alfa Romeo Giulietta Gold Portfolio 1954-1965
Alfa Romeo Giulia Berlinas 1962-1976
Alfa Romeo Giulia Coupés 1963-1976
Alfa Romeo Giulia Coupés Gold Port. 1963-1976
Alfa Romeo Spider 1966-1990
Alfa Romeo Spider Gold Portfolio 1966-1991
Alfa Romeo Alfasud 1972-1984
Alfa Romeo Alfetta Gold Portfolio 1972-1987
Alfa Romeo Alfetta GTV6 1980-1986
Allard Gold Portfolio 1937-1959
Alvis Gold Portfolio 1919-1967
AMX & Javelin Muscle Portfolio 1968-1974
Armstrong Siddeley Gold Portfolio 1945-1960
Aston Martin Gold Portfolio 1948-1971
Aston Martin Gold Portfolio 1972-1985
Aston Martin Gold Portfolio 1985-1995
Audi Quattro Gold Portfolio 1980-1991
Austin A30 & A35 1951-1962
Austin Healey 100 & 100/6 Gold Port. 1952-1959
Austin Healey 3000 Gold Portfolio 1959-1967
Austin Healey Sprite Gold Portfolio 1958-1971
Barracuda Muscle Portfolio 1964-1974
BMW 1600 Collection No.1 1966-1981
BMW 2002 Gold Portfolio 1968-1976
BMW 316, 318, 320 (4 cyl.) Gold Port. 1975-1990
BMW 320, 323, 325 (6 cyl.) Gold Port. 1977-1990
BMW M Series Performance Portfolio 1976-1993
BMW 5 Series Gold Portfolio 1981-1987
BMW 6 Series Gold Portfolio 1976-1989
Bricklin Gold Portfolio 1974-1975
Bristol Cars Gold Portfolio 1946-1992
Buick Automobiles 1947-1960
Buick Muscle Cars 1965-1970
Cadillac Allanté 1986-1993
Cadillac Automobiles 1949-1959
Cadillac Automobiles 1960-1969
Caprice 1965-1976 ☆ Limited Edition
Charger Muscle Portfolio 1966-1974
Checker ☆ Limited Edition
Chevrolet 1955-1957
Impala & SS Muscle Portfolio 1958-1972
Chevrolet Corvair 1959-1969
Chevy II & Nova SS Muscle Portfolio 1962-1974
Chevy El Camino & SS 1959-1987
Chevelle & SS Muscle Portfolio 1964-1972
Chevrolet Muscle Cars 1966-1971
Chevy Blazer 1969-1981
Chevrolet Corvette Gold Portfolio 1953-1962
Chevrolet Corvette Sting Ray Gold Port. 1963-1967
Chevrolet Corvette Gold Portfolio 1968-1977
High Performance Corvettes 1983-1989
Camaro Muscle Portfolio 1967-1973
Chevrolet Camaro & Z28 1973-1981
High Performance Camaros 1982-1988
Chrysler 300 Gold Portfolio 1955-1970
Chrysler Valiant 1960-1962
Citroen Traction Avant Gold Portfolio 1934-1957
Citroen 2CV Gold Portfolio 1948-1989
Citroen DS & ID 1955-1975
Citroen DS & ID Gold Portfolio 1955-1975
Citroen SM 1970-1975
Cobras & Replicas 1962-1983
Shelby Cobra Gold Portfolio 1962-1969
Cobras & Cobra Replicas Gold Portfolio 1962-1989
Cunningham Automobiles 1951-1955
Daimler SP250 Sports & V-8 250 Saloon Gold P. 1959-1969
Datsun Roadsters 1962-1971
Datsun 240Z 1970-1973
Datsun 280Z & ZX 1975-1983
DeLorean Gold Portfolio 1977-1995
Dodge Muscle Cars 1967-1970
Dodge Viper on the Road
Edsel 1957-1960 ☆ Limited Edition
ERA Gold Portfolio 1934-1994
Excalibur Collection No.1 1952-1981
Facel Vega 1954-1964
Ferrari Dino 1965-1974
Ferrari Dino 308 & Mondial Gold Portfolio1974-1985
Ferrari 328 • 348 • Mondial Gold Portfolio 1986-1994
Fiat 500 Gold Portfolio 1936-1972
Fiat 600 & 850 Gold Portfolio 1955-1972
Fiat Pininfarina 124 & 2000 Spider 1968-1985
Fiat X1/9 Gold Portfolio1973-1989
Fiat Abarth Performance Portfolio 1972-1987
Ford Consul, Zephyr, Zodiac Mk.I & II 1950-1962
Ford Zephyr, Zodiac, Executive, Mk.III & Mk.IV 1962-1971
Ford Cortina 1600E & GT 1967-1970
High Performance Capris Gold Portfolio 1969-1987
Capri Muscle Portfolio 1974-1987
High Performance Fiestas 1979-1991
High Performance Escorts Mk.I 1968-1974
High Performance Escorts Mk.II 1975-1980
High Performance Escorts 1980-1985
High Performance Escorts 1985-1990
High Performance Sierras & Merkurs
Gold Portfolio 1983-1990
Ford Automobiles 1949-1959
Ford Fairlane 1955-1970
Ford Ranchero 1957-1959
Ford Thunderbird 1955-1957
Ford Thunderbird 1958-1963
Ford GT40 Gold Portfolio 1964-1987
Ford Bronco 1966-1977
Ford Bronco 1978-1988
Goggomobil ☆ Limited Edition
Holden 1948-1962
Honda CRX 1983-1987
Imperial 1955-1970 ☆ Limited Edition
International Scout Gold Portfolio 1961-1980
Isetta 1953-1964
Iso & Bizzarrini Gold Portfolio 1962-1974
Kaiser • Frazer 1946-1955 ☆ Limited Edition
Jaguar and SS Gold Portfolio 1931-1951
Jaguar XK120, 140, 150 Gold Port. 1948-1960
Jaguar Mk.VII, VIII, IX, X, 420 Gold Port. 1950-1970
Jaguar Mk.1 & Mk.2 Gold Portfolio 1959-1969
Jaguar C-Type & D-Type ☆ Limited Edition
Jaguar E-Type Gold Portfolio 1961-1971
Jaguar E-Type V-12 1971-1975
Jaguar S-Type & 420 ☆ Limited Edition
Jaguar XJ12, XJ5.3, V12 Gold Portfolio 1972-1990
Jaguar XJ6 Series I & II Gold Portfolio 1968-1979
Jaguar XJ6 Series III Perf. Portfolio 1979-1986
Jaguar XJ6 Gold Portfolio 1986-1994
Jaguar XJS Gold Portfolio 1975-1988
Jaguar XJS Gold Portfolio 1988-1995
Jeep CJ5 & CJ6 1960-1976
Jeep CJ5 & CJ7 1976-1986
Jensen Cars 1946-1967
Jensen Cars 1967-1979
Jensen Interceptor Gold Portfolio 1966-1986
Jensen Healey 1972-1976
Lagonda Gold Portfolio 1919-1964
Lamborghini Countach & Urraco 1974-1980
Lamborghini Countach & Jalpa 1980-1985
Lancia Aurelia & Flaminia Gold Portfolio 1950-1970
Lancia Fulvia Gold Portfolio 1963-1976
Lancia Beta Gold Portfolio 1972-1984
Lancia Delta Gold Portfolio 1979-1994
Lancia Stratos 1972-1985
Land Rover Series I 1948-1958
Land Rover Series II & IIa 1958-1971
Land Rover Series III 1971-1985
Land Rover 90 110 Defender Gold Portfolio 1983-1994
Land Rover Discovery 1989-1994
Land Rover Story Part One 1948-1971
Lincoln Gold Portfolio 1949-1960
Lincoln Continental 1961-1969
Lincoln Continental 1969-1976
Lotus Sports Racers Gold Portfolio 1953-1965
Lotus Seven Gold Portfolio 1957-1974
Lotus Caterham Seven Gold Portfolio 1974-1995
Lotus Elan Gold Portfolio 1962-1974
Lotus Elan Collection No. 2 1963-1972
Lotus Elan & SE 1989-1992
Lotus Europa Gold Portfolio 1966-1975
Lotus Elite & Eclat 1974-1982
Lotus Turbo Esprit 1980-1986
Marcos Cars 1960-1988
Maserati 1965-1975
Matra 1965-1983 ☆ Limited Edition
Mazda Miata MX-5 Performance Portfolio 1989-1996
Mazda RX-7 Gold Portfolio 1978-1991
Mercedes 190 & 300 SL 1954-1963
Mercedes G Wagen 1981-1994
Mercedes S & 600 1965-1972
Mercedes S Class 1972-1979
Mercedes 230 • 250 • 280SL Gold Portfolio 1963-1971
Mercedes SLs & SLCs Gold Portfolio 1971-1989
Mercedes SLs Performance Portfolio 1989-1994
Mercury Muscle Cars 1966-1971
Messerschmitt Gold Portfolio 1954-1964
MG Gold Portfolio 1929-1939
MG TA & TC Gold Portfolio 1936-1949
MG TD & TF Gold Portfolio 1949-1955
MGA & Twin Cam Gold Portfolio 1955-1962
MG Midget Gold Portfolio 1961-1979
MGB Roadsters 1962-1980
MGB MGC & V8 Gold Portfolio 1962-1980
MGB GT 1965-1980
MGC & MGB GT V8 ☆ Limited Edition
MG Y-Type & Magnette ZA/ZB ☆ Limited Edition
Mini Gold Portfolio 1959-1969
Mini Gold Portfolio 1969-1980
High Performance Minis Gold Portfolio 1960-1973
Mini Cooper Gold Portfolio 1961-1971
Mini Moke Gold Portfolio 1964-1994
Mopar Muscle Cars 1964-1967
Morgan Three-Wheeler Gold Portfolio 1910-1952
Morgan Plus 4 & Four 4 Gold Portfolio. 1936-1967
Morgan Cars 1960-1970
Morgan Cars Gold Portfolio 1968-1989
Morris Minor Collection No. 1 1948-1980
Shelby Mustang Muscle Portfolio 1965-1970
High Performance Mustang IIs 1974-1978
High Performance Mustangs 1982-1988
Nash & Nash-Healey 1949-1957 ☆ Limited Edition
Nash-Austin Metropolitan Gold Portfolio 1954-1962
Oldsmobile Automobiles 1955-1963
Oldsmobile Muscle Cars 1964-1971
Oldsmobile Toronado 1966-1978
Opel GT Gold Portfolio 1968-1973
Opel Manta 1970-1975 ☆ Limited Edition
Packard Gold Portfolio 1946-1958
Pantera Gold Portfolio 1970-1989
Panther Gold Portfolio 1972-1990
Plymouth Muscle Cars 1966-1971
Pontiac Tempest & GTO 1961-1965
Pontiac Muscle Cars 1966-1972
Pontiac Firebird & Trans-Am 1973-1981
High Performance Firebirds 1982-1988
Pontiac Fiero 1984-1988
Porsche 356 Gold Portfolio 1953-1965
Porsche 911 1965-1969
Porsche 911 1970-1972
Porsche 911 1973-1977
Porsche 911 SC & Turbo Gold Portfolio 1978-1983
Porsche 911 Carrera & Turbo Gold Port. 1984-1989
Porsche 924 Gold Portfolio 1975-1988
Porsche 928 Performance Portfolio 1977-1994
Porsche 944 Gold Portfolio 1981-1991
Range Rover Gold Portfolio 1970-1985
Range Rover Gold Portfolio 1986-1995
Reliant Scimitar 1964-1986
Renault Alpine Gold Portfolio 1958-1994
Riley Gold Portfolio 1924-1939
Rolls Royce Silver Cloud & Bentley 'S' Series
Gold Portfolio 1955-1965
Rolls Royce Silver Shadow Gold Port. 1965-1980
Rolls Royce & Bentley Gold Port. 1980-1989
Rover P4 1949-1959
Rover P4 1955-1964
Rover 3 & 3.5 Litre Gold Portfolio 1958-1973
Rover 2000 & 2200 1963-1977
Rover 3500 1968-1977
Rover 3500 & Vitesse 1976-1986
Saab Sonett Collection No.1 1966-1974
Saab Turbo 1976-1983
Studebaker Gold Portfolio 1947-1966
Studebaker Hawks & Larks 1956-1963
Avanti 1962-1990
Sunbeam Tiger & Alpine Gold Portfolio. 1959-1967
Toyota Land Cruiser 1956-1984
Triumph Dolomite Sprint ☆ Limited Edition
Triumph TR2 & TR3 Gold Portfolio 1952-1961
Triumph TR4, TR5, TR250 1961-1968
Triumph TR6 Gold Portfolio 1969-1976
Triumph TR7 & TR8 Gold Portfolio 1975-1982
Triumph Herald 1959-1971
Triumph Vitesse 1962-1971
Triumph Spitfire Gold Portfolio 1962-1980
Triumph 2000, 2.5, 2500 1963-1977
Triumph GT6 Gold Portfolio 1966-1974
Triumph Stag Gold Portfolio 1970-1977
TVR Gold Portfolio 1959-1986
TVR Performance Portfolio 1986-1994
VW Beetle Gold Portfolio 1935-1967
VW Beetle Gold Portfolio 1968-1991
VW Beetle Collection No.1 1970-1982
VW Karmann Ghia 1955-1982
VW Bus, Camper, Van 1954-1967
VW Bus, Camper, Van 1968-1979
VW Bus, Camper, Van 1979-1989
VW Scirocco 1974-1981
VW Golf GTI 1976-1986
Volvo PV444 & PV544 1945-1965
Volvo Amazon-120 Gold Portfolio 1956-1970
Volvo 1800 Gold Portfolio 1960-1973
Volvo 140 & 160 Series Gold Portfolio 1966-1975
Westfield ☆ Limited Edition

Forty Years of Selling Volvo

BROOKLANDS ROAD & TRACK SERIES

Road & Track on Alfa Romeo 1964-1970
Road & Track on Alfa Romeo 1971-1976
Road & Track on Aston Martin 1962-1990
R & T on Auburn Cord and Duesenburg 1952-84
Road & Track on Audi & Auto Union 1952-1980
Road & Track on Audi & Auto Union 1980-1986
Road & Track on Austin Healey 1953-1970
Road & Track on BMW Cars 1966-1974
Road & Track on BMW Cars 1975-1978
Road & Track on BMW Cars 1979-1983
R & T on Cobra, Shelby & Ford GT40 1962-1992
Road & Track on Corvette 1953-1967
Road & Track on Corvette 1968-1982
Road & Track on Corvette 1982-1986
Road & Track on Corvette 1986-1990
Road & Track on Ferrari 1975-1981
Road & Track on Ferrari 1981-1984
Road & Track on Ferrari 1984-1988
Road & Track on Fiat Sports Cars 1968-1987
Road & Track on Jaguar 1950-1960
Road & Track on Jaguar 1961-1968
Road & Track on Jaguar 1968-1974
Road & Track on Jaguar 1974-1982
Road & Track on Jaguar 1983-1989
Road & Track on Lamborghini 1964-1985
Road & Track on Lotus 1972-1981
Road & Track on Maserati 1975-1983
R & T on Mazda RX-7 & MX-5 Miata 1986-1991
Road & Track on Mercedes 1952-1962
Road & Track on Mercedes 1963-1970
Road & Track on Mercedes 1971-1979
Road & Track on Mercedes 1980-1987
Road & Track on MG Sports Cars 1949-1961
Road & Track on MG Sports Cars 1962-1980
Road & Track on Mustang 1964-1977
R & T on Nissan 300-ZX & Turbo 1984-1989
Road & Track on Pontiac 1960-1983
Road & Track on Porsche 1951-1967
Road & Track on Porsche 1968-1971
Road & Track on Porsche 1972-1975
Road & Track on Porsche 1975-1978
Road & Track on Porsche 1985-1988
R & T on Rolls Royce & Bentley 1950-1965
R & T on Rolls Royce & Bentley 1966-1984
Road & Track on Saab 1972-1992
R & T on Toyota Sports & GT Cars 1966-1984
R & T on Triumph Sports Cars 1953-1967
R & T on Triumph Sports Cars 1967-1974
R & T on Triumph Sports Cars 1974-1982
Road & Track on Volkswagen 1951-1968
Road & Track on Volkswagen 1968-1978
Road & Track on Volkswagen 1978-1985
Road & Track on Volvo 1957-1974
Road & Track on Volvo 1977-1994
R & T - Henry Manney at Large & Abroad
R & T - Peter Egan's "Side Glances"
R & T - Peter Egan "At Large"

BROOKLANDS CAR AND DRIVER SERIES

Car and Driver on BMW 1955-1977
Car and Driver on BMW 1977-1985
C and D on Cobra, Shelby & Ford GT40 1963-84
Car and Driver on Corvette 1978-1982
Car and Driver on Corvette 1983-1988
C and D on Datsun Z 1600 & 2000 1966-1984
Car and Driver on Ferrari 1955-1962
Car and Driver on Ferrari 1963-1975
Car and Driver on Ferrari 1976-1983
Car and Driver on Mopar 1956-1967
Car and Driver on Mopar 1968-1975
Car and Driver on Mustang 1964-1972
Car and Driver on Pontiac 1961-1975
Car and Driver on Porsche 1955-1962
Car and Driver on Porsche 1963-1970
Car and Driver on Porsche 1970-1976
Car and Driver on Porsche 1977-1981
Car and Driver on Porsche 1982-1986
Car and Driver on Saab 1956-1985
Car and Driver on Volvo 1955-1986

BROOKLANDS PRACTICAL CLASSICS SERIES

PC on Austin A40 Restoration
PC on Land Rover Restoration
PC on Metalworking in Restoration
PC on Midget/Sprite Restoration
PC on MGB Restoration
PC on Sunbeam Rapier Restoration
PC on Triumph Herald/Vitesse
PC on Spitfire Restoration
PC on Beetle Restoration
PC on 1930s Car Restoration

BROOKLANDS HOT ROD 'MUSCLECAR & HI-PO ENGINES' SERIES

Chevy 265 & 283
Chevy 302 & 327
Chevy 348 & 409
Chevy 350 & 400
Chevy 396 & 427
Chevy 454 thru 512
Chrysler Hemi
Chrysler 273, 318, 340 & 360
Chrysler 361, 383, 400, 413, 426, 440
Ford 289, 302, Boss 302 & 351W
Ford 351C & Boss 351
Ford Big Block

BROOKLANDS RESTORATION SERIES

Auto Restoration Tips & Techniques
Basic Bodywork Tips & Techniques
Camaro Restoration Tips & Techniques
Chevrolet High Performance Tips & Techniques
Chevy Engine Swapping Tips & Techniques
Chevy-GMC Pickup Repair
Chrysler Engine Swapping Tips & Techniques
Engine Swapping Tips & Techniques
Ford Pickup Repair
How to Build a Street Rod
Land Rover Restoration Tips & Techniques
MG 'T' Series Restoration Guide
MGA Restoration Guide
Mustang Restoration Tips & Techniques
Performance Tuning - Chevrolets of the '60's
Performance Tuning - Pontiacs of the '60's

MOTORCYCLING

BROOKLANDS ROAD TEST SERIES

AJS & Matchless Gold Portfolio 1945-1966
BSA Twins A7 & A10 Gold Portfolio 1946-1962
BSA Twins A50 & A65 Gold Portfolio 1962-1973
BMW Motorcycles Gold Portfolio 1950-1971
BMW Motorcycles Gold Portfolio 1971-1976
Ducati Gold Portfolio 1960-1974
Ducati Gold Portfolio 1974-1978
Ducati Gold Portfolio 1978-1982
Laverda Gold Portfolio 1967-1977
Moto Guzzi Gold Portfolio 1949-1973
Norton Commando Gold Portfolio 1968-1977
Triumph Bonneville Gold Portfolio 1959-1983

BROOKLANDS CYCLE WORLD SERIES

Cycle World on BMW 1974-1980
Cycle World on BMW 1981-1986
Cycle World on Ducati 1982-1991
Cycle World on Harley-Davidson 1962-1968
Cycle World on Harley-Davidson 1978-1983
Cycle World on Harley-Davidson 1983-1987
Cycle World on Harley-Davidson 1987-1990
Cycle World on Harley-Davidson 1990-1992
Cycle World on Honda 1962-1967
Cycle World on Honda 1968-1971
Cycle World on Honda 1971-1974
Cycle World on Husqvarna 1966-1976
Cycle World on Husqvarna 1977-1984
Cycle World on Kawasaki 1966-1971
Cycle World on Kawasaki Off-Road Bikes 1972-1979
Cycle World on Kawasaki Street Bikes 1972-1976
Cycle World on Norton 1962-1971
Cycle World on Suzuki 1962-1970
Cycle World on Suzuki Off-Road Bikes 1971-1976
Cycle World on Suzuki Street Bikes 1971-1976
Cycle World on Triumph 1967-1972
Cycle World on Yamaha 1962-1969
Cycle World on Yamaha Off-Road Bikes 1970-1974
Cycle World on Yamaha Street Bikes 1970-1974

MILITARY

BROOKLANDS MILITARY VEHICLES SERIES

Allied Military Vehicles No.2 1941-1946
Complete WW2 Military Jeep Manual
Dodge Military Vehicles No.1 1940-1945
Hail To The Jeep
Military & Civilian Amphibians 1940-1990
Off Road Jeeps: Civ. & Mil. 1944-1971
US Military Vehicles 1941-1945
US Army Military Vehicles WW2-TM9-2800
VW Kubelwagen Military Portfolio 1940-1990
WW 2 Jeep Military Portfolio 1941-1945

RACING

Le Mans - The Jaguar Years - 1949-1957
Le Mans - The Ferrari Years - 1958-1965
Le Mans - The Ford & Matra Years - 1966-1974

2117

Brooklands Books

CONTENTS

ACKNOWLEDGEMENTS

The addition of a Moto Guzzi title to Brooklands Books' Gold Portfolio series underlines the impact made by this Italian manufacturer when their road-going models were imported into the UK during the early post-war years. By selecting the more authoritative magazine features and collating them within a single volume, it has been possible to give a comprehensive across-the-range coverage. Even the exotic vee eight racing model is included, a machine that created a sensation during the 1957 Golden Jubilee TT in the Isle of Man.

Many famous names have been associated with Moto Guzzi, foremost amongst them Stanley Woods, the legendary Irishman who enjoyed an incredible number of racing successes. It was not only in racing, however, that they made their mark. Their road-going models, finished in Italian racing red, appeared to British enthusiasts to offer something quite different from that available on the home market.

It would not have been possible to present all the information contained within this book had it not been for the generous assistance given by the world's leading magazine publishers. Brooklands Books would like to express their appreciation to each of the following, who so willingly gave permission for the use of their copyright material: *Cycle, Cycle Buyers Guide, Cycle Guide, Cycle World, The Motor Cycle, Motorcycle Mechanics, Motorcycle Sport, Motor Cycling, Motorcyclist, Motorcyclist Illustrated* and *Two Wheels.*

Our thanks also go to Dave Minton of Classic Bike Guide for supplying the colour photograph of a Moto Guzzi Ambassador that appears on our front cover.

R.M. Clarke

Moto Guzzi began making motorcycles in 1921 at their factory by Lake Como and were known in the UK, almost exclusively through their racing successes, until the early post-war years. It was then that the importation of their road-going models commenced, to add a touch of colour to the otherwise somewhat drab motorcycling scene. The models that attracted the most attention were the Unit-construction 175cc Lodola, later increased in capacity to 235cc, and the 110cc Zigolo two-stroke, also built along similar lines. Their attractive appearance was enhanced by their finish in Italian racing red.

Although its cost in the UK put it out of contention with its contemporaries, the 192cc Galletto scooter was also worthy of attention. Virtually a motorcycle enclosed within a metal skin, those who had the good fortune to ride one considered it one of the best scooters ever made.

The Moto Guzzi that drew most attention was their vee-eight racing model, a truly exotic machine that was entered in the Golden Jubilee TT of 1957, to be ridden by Dickie Dale. Nothing quite like it had been seen before and it was little short of amazing how the engine unit with its six-speed gearbox could be slotted into a racing frame and cycle parts. Capable of producing 62bhp at 12,000rpm, it had a maximum speed of 160mph. Sadly, the best it could do was to finish fourth, with engine problems, on what amounted to its first serious outing.

Until then, when it came to making the larger capacity models Moto Guzzi had always shown their preference for wide angle vee twins. When they returned to making larger capacity road models they still used the same basic engine layout, although the angle between the cylinders was reduced to 90 degrees and the engine was mounted transversely in the frame, with the use of shaft final drive. Many derivatives of the 750cc and 850cc twins followed, including a 949cc version with automatic transmission in 1975.

Jeff Clew

ROAD TESTS OF NEW MODELS

246 c.c. Overhead-valve Moto Guzzi

ONE of the most famous makes on the European continent in virtue of consistent racing successes is the Guzzi which this year won the manufacturers' World's Championship in the 250 c.c. class. The production 250 c.c. model intended for the sporting tourist is the "Airone Sport" which, in general layout, such as the horizontal single cylinder, unit-construction of engine and gear box, and pivot-action rear suspension follows characteristic Guzzi design of the racing singles.

Moto Guzzi "Airone Sport"—a lively two-fifty with a racing pedigree

Unquestionably, this sports machine bears the influence of a racing pedigree. If one attribute had to be singled out among the many with which this machine is endowed, it would be its suitability for maintaining high average speeds over long distances. The engine is tireless and belies its comparatively small size; steering and road-holding in the higher speed ranges are faultless.

As will be noticed in the performance data table, the mean maximum speed is good, but not remarkable for a 250 c.c. machine, though here it might be noted that under only slightly favourable conditions a maximum of well over 70 m.p.h. can be achieved. The top gear ratio of 6 to 1 is on the high side, which in itself is an attractive point, though it does militate against the mean maximum figure.

But out-and-out speed is unimportant. Much more to the point is the fact that speeds higher than 60 m.p.h. can be held, it seemed indefinitely, as often as road conditions allow. It is a machine which can be driven on full throttle for mile after mile with the certain knowledge that the engine will not tire. Under all high-speed conditions the engine produces its power smoothly, effortlessly and, it should be noted, economically. No matter how hard the driving conditions, the engine remains noticeably cool; in fact, on some occasions when temperatures were low, the impression was gained that the result of a light alloy cylinder head and barrel with the head forward and receiving the major effect of the air stream, was over-cooling—this in spite of the fact that the carburettor settings gave a mixture slightly on the lean side.

Acceleration is sprightly, and although the engine has very good pulling powers, it very freely reached high revolutions in the indirect gears. Valve float was not too easy to identify and, when it had started, the revs would continue to mount.

Both front and rear suspensions are near perfect for high speeds. When travelling fast on average main roads, the Guzzi was comfortable to a marked degree, and with a 12-stone rider there seemed to be no advantage in restricting the rear-springing by means of the adjustable friction dampers provided. The suspension gives that "tight," precise sensation more usually associated with the road-holding of racing machines. This is achieved by springing which is rather too hard to give maximum comfort at low speeds. For instance, at around 15 m.p.h. it was possible to traverse minor road bumps and find that the suspension was absorbing few of the shocks. In particular, the front fork characteristics contrasted sharply with the soft static-load position springing of the average telescopic fork of to-day.

Steering is first class in all circumstances, and it was never necessary to employ the steering damper. The steering is light and positive; it consistently retains these features from a slow walking pace up to maximum speed and is unaffected by road conditions. Handleable in the full sense of the term, the Guzzi is a sheer delight when the mood is one for cornering zestfully on an accurate line. The impression gained is that the machine is lighter than the scales indicate and stability is markedly good on wet, slippery road surfaces.

Whether cold or not, the engine fired readily provided that, under cold conditions, the carburettor was flooded and the air control closed. The kick-starter is on the left-hand side and did not seem so easy to operate as a right-hand side pedal; this should not, however, be regarded as a criticism, since many years of experience with the orthodox British layout may be expected to have resulted in prejudice as well as the acquisition of knack. Engine idling was always regular and reliable, and if the twistgrip throttle was set carefully and the ignition control retarded, a very slow tickover could be obtained. The carburettor is not fitted with a throttle stop, and lost

Unit-construction of engine and gear box with gear primary drive is employed

Robust Italian Machine Capable of Withstanding Very Hard Driving : Remarkably Light Petrol Consumption

motion in the rather slow-action twistgrip, and in the cable, did not allow an accurate setting to give slow idling with the grip rolled right back.

The clutch freed perfectly and took up the drive smoothly; operation was inclined to be heavy. Gear changes with the rocking pedal could be accomplished quickly and cleanly at all speeds, but this control also required more pressure than expected; for example, although downward changes could be made with the toe as on the majority of British machines, the effort required was considerable and it was easier to employ the heel pedal which involved moving the foot back from its normal position on the rest. The gear box was quiet in the first and second ratios, but slightly audible in third. So far as could be ascertained, the gear primary drive was noiseless. Transmission lacked the slight flexibility usually associated with full chain drive incorporating a good shock-absorber.

Deep-tone Exhaust

Operation of the valve gear was audible, but in all other respects the engine was quiet. The silencer gives a deep tone which, at low speeds, is perhaps rather loud by British single-cylinder standards. However, the silencer is equally effective when the engine is revving fast; at highish speeds exhaust noise is not obtrusive and is commendably free from high-pitch frequencies.

Brakes are of large diameter and, towards the end of the mileage of the test, became smooth and powerful. When the machine was received it had covered fewer than 100 miles, and the front brake required heavy lever pressure before it was effective. Later this brake improved markedly, and a second set of stopping-distance tests was carried out which showed a decrease of nearly 3ft over the earlier figure achieved; the shorter stopping distance is given in the information panel. Rear-brake operation is by heel on the left side; to apply the brake the foot had to be moved back on the footrest and it was difficult to achieve the delicacy usual with toe operation.

An impression of "compactness" is given by the riding position which is very well suited to riders of varying stature. A range of adjustment is provided at the handlebar and the footrests. The mudguards have deep valances, and legshields are fitted; the protection given proved highly effective under dirty road conditions. The beam from the head lamp provides an excellent driving light; when the out-of-focus filament is brought into use by the handlebar switch there is a distinct cut-off of light above the horizontal plane which satisfies the most stringent anti-dazzle criterion, yet the light on the road is adequate. No praise can be too high for the head lamp fitted.

A central stand with very large roll-on feet is fitted. Next to no effort was required to bring the stand into use—foot pressure on the left-side of the stand was sufficient. There is no front stand, and considerable weight on the rear was necessary to pivot the machine about the central stand and bring the front wheel clear of the ground.

The Guzzi design is strictly functional and by British standards might be regarded as untidy. For example, the kick-starter and foot-change mechanisms are external, and no effort is made to conceal linkages and pipes or to harmonize the machine's lines. However, of its robustness and suitability for sustained hard work there can be no doubt.

A neat instrument panel housing speedometer and clock is fitted. The speedometer, calibrated in kilometres, registered approximately 15 per cent fast

Information Panel

SPECIFICATION

ENGINE : 246 c.c. (70 x 64 mm) single-cylinder push-rod operated o.h.v., with four-speed gear box in unit. Hairpin valve-springs ; valve gear fully enclosed, light-alloy cylinder head and cylinder barrel. Dry-sump lubrication ; tank capacity, $3\frac{1}{2}$ pints.

CARBURETTOR : Dellorto ; twistgrip throttle and handlebar air lever.

TRANSMISSION : Guzzi four-speed gear box with positive stop foot control. Gear ratios : Top, 6.0 to 1. Third, 8.05 to 1. Second, 10.7 to 1. Bottom, 15.8 to 1. Primary drive by gears. Secondary drive by $\frac{1}{2} \times \frac{1}{4}$in chain with guard over top run. R.p.m. at 30 m.p.h. in top gear, 2,373 approximately.

IGNITION AND LIGHTING : Marelli magneto with manual timing control by handlebar lever. Marelli dynamo ; $6\frac{1}{2}$in head lamp with sealed beam light unit.

FUEL CAPACITY : $2\frac{1}{2}$ gallons.

TYRES : Pirelli ; front ribbed, rear studded ; both 3.00 x 19in on light-alloy rims.

BRAKES : 8in diameter.

SUSPENSION : Telescopic front fork controlled by springs and hydraulic damping ; pivot-action rear-springing with adjustable friction dampers.

WHEELBASE : 54in. Ground clearance, $5\frac{1}{2}$in unladen.

SADDLE : Radaelli. Unladen height, 29in.

WEIGHT : 343lb with full oil-tank, approx. one gallon of fuel, and fully equipped.

PRICE : Will be available shortly

ROAD TAX : £1 17s. 6d. a year (10s. 4d. a quarter). Half rate if used only on standard ration.

MAKERS : Moto Guzzi S.A., Mandello del Lario, Como, Italy. British agents, Bob Foster, 472-4, Ashley Road, Parkstone, Dorset.

DESCRIPTION : *The Motor Cycle* dated October 7, 1948.

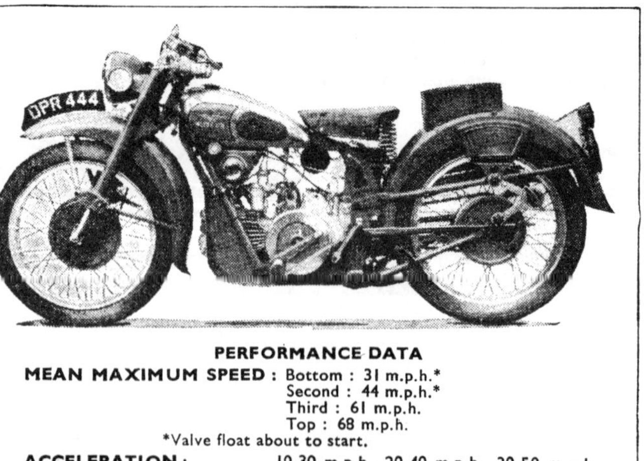

PERFORMANCE DATA

MEAN MAXIMUM SPEED : Bottom : 31 m.p.h.*
Second : 44 m.p.h.*
Third : 61 m.p.h.
Top : 68 m.p.h.

*Valve float about to start.

ACCELERATION :

	10-30 m.p.h.,	20-40 m.p.h.,	30-50 m.p.h.
Bottom	3.8 secs	—	—
Second	4.4 secs	4 secs	—
Third	7.2 secs	6.8 secs	7.8 secs
Top	—	11.4 secs	11.2 secs

Speed at end of quarter-mile from rest : 63 m.p.h.
Time to cover standing quarter-mile : 20.4 secs.

PETROL CONSUMPTION : At 30 m.p.h., 124 m.p.g. At 40 m.p.h., 109 m.p.g. At 50 m.p.h., 96 m.p.g. At 60 m.p.h., 79 m.p.g.

BRAKING : From 30 m.p.h. to rest, 29ft (surface, dry tar macadam).

TURNING CIRCLE : 13ft 6in.

MINIMUM NON-SNATCH SPEED : 13 m.p.h. in top gear.

WEIGHT PER C.C. : 1.4lb.

500 cc single-cylinder racing Moto Guzzi. This model is available to ordinary racing riders desiring to campaign on the Continent and at the Isle of Man.

Unusual type of all-enclosed machine equipped with a spare wheel complete. Engine is 150 cc, single cylinder, overhead valve, four stroke.

LEFT—Single-cylinder Moto Guzzi engine-gearbox assembly, viewed from right side, showing clean lines, compactness, and accessibility of components. Italian designers exercise great ingenuity and apply daring imagination to their products. The Moto Guzzi firm pioneered the horizontal single-cylinder design of engine. Theory suggested that single-cylinder power impulses could be more effectively dampened if the direction of the impulses were in line with the wheelbase, not at right angles to it. Theory was tested and proved sound. Moto Guzzi have featured this example of imaginative engineering throughout their full line. Horizontal cooling fins and down-draft intake manifolding are interesting points. This particular photo depicts an engine designed for use in 3-wheel package delivery van, employs shaft drive. Castings are light alloy wherever possible.

ITALIAN MOTO GUZZI MOTORCYCLES

"A terribly grotesque machine" is the quick appraisal of the twin cylinder 120 degree Moto Guzzi racing 500 cc bike. Built for pure functionalism. Enormous front brake, contoured gas tank, swinging arm rear suspension are notable features.

One of the most popular Moto Guzzi designs—the 250 cc road machine so beloved by Continental riders. Features full road equipment for touring, heel operated rear brake, roll-up type stand, plunger forks, outside flywheel, enclosed overhead valves, full loop-type tubular frame.

Final assembly line of 98 cc and 125 cc models. Much hand work is evident on all Moto Guzzi machines, extreme care in final assembly assures ultimate rider acceptance and approval.

MODERN ENGINES

500 c.c. Moto Guzzi

A Highly Successful Racing Unit, Evolved from a Standard Engine

By FERGUS ANDERSON

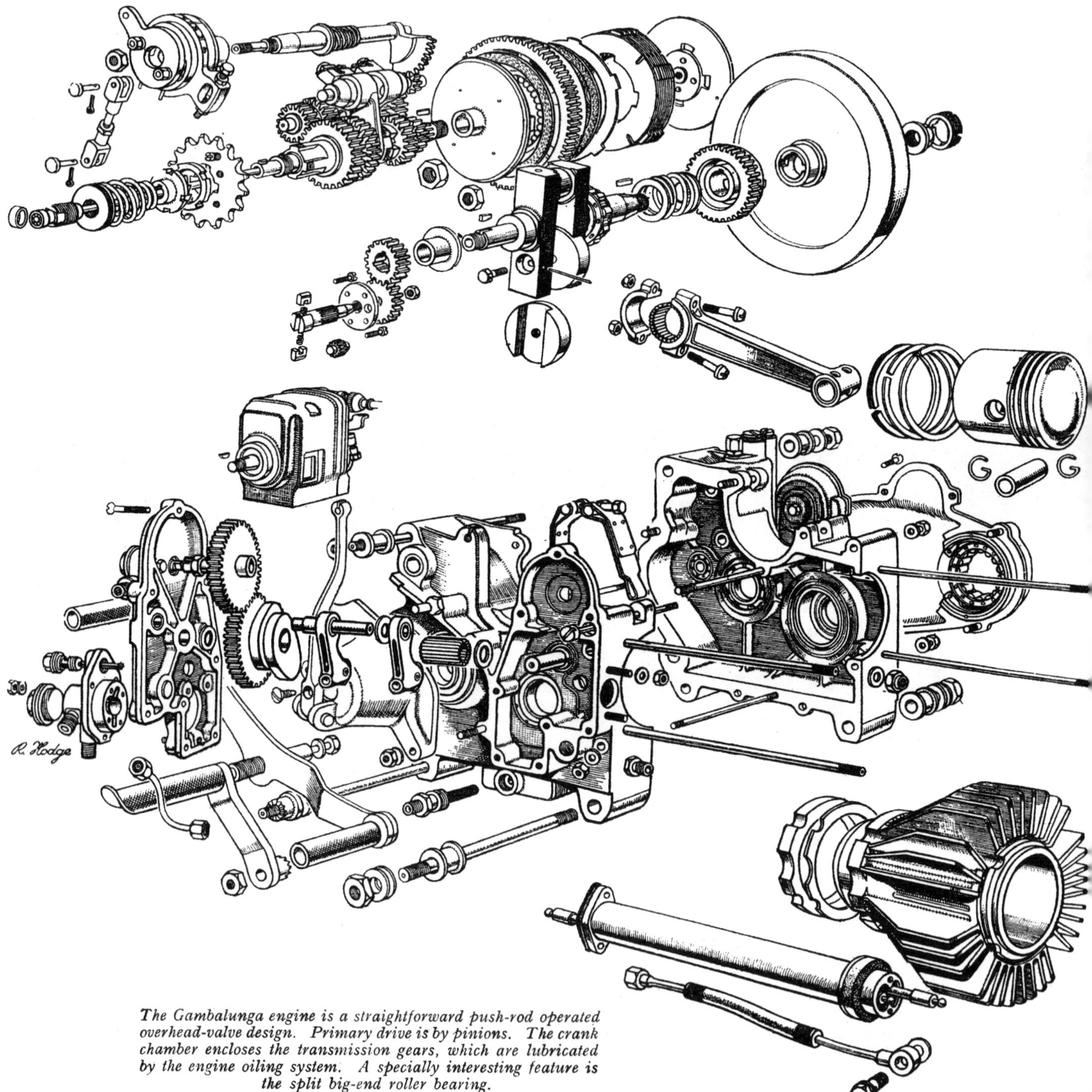

The Gambalunga engine is a straightforward push-rod operated overhead-valve design. Primary drive is by pinions. The crank chamber encloses the transmission gears, which are lubricated by the engine oiling system. A specially interesting feature is the split big-end roller bearing.

Gambalunga

Which Bristles With Unconventionalities

BEFORE the war Moto Guzzi were making a super-sports machine called the Condor (later, Dondolino), and it was on such a machine that one of the factory testers, Ferdinando Balzarotti, began to show that he was likely to develop into a first-class rider. When, after the war, racing re-started in Italy, the Guzzi racing designer, Giulio Carcano, was asked to see whether he could do something towards providing Balzarotti with a rather better machine. The "something" turned out to the "Gambalunga."

Habitually a race goes round and round, eventually arriving back where it started. The evolution of the Gambalunga engine took the same path for, starting with the bore and stroke of the Condor (88 × 82mm) the first step taken was to reduce the bore and lengthen the stroke to give dimensions of 84 × 90mm. Hence the name, for Gambalunga, translated means "Long Leg." The latest engines are called "Gambalunga—Tipo Faenza" (Gambalunga—Faenza type) for in the Italian Grand Prix at Faenza in 1948, Lorenzetti used an engine with the original bore and stroke and these dimensions are now used.

In point of performance, there is little to choose between the engines, but the Tipo Faenza develops its power at lower revolutions, with consequential reduced stresses on the engine. The accompanying drawing shows the long-stroke unit (the thick "compression plate" under the cylinder is to compensate for the increase in stroke) and, except for obvious compensating changes, there are no structural differences between the engines. The same cylinder head is used, though, in the later engine, it has to be chamfered off to match up with the bore of the cylinder.

The Elektron crankcase contains the crankshaft and gear assembly in a common chamber. The fact that the whole of the crankcase is acting as an oil radiator makes it possible to fair in the underneath of the engine without overheating. The forged-steel crankshaft (VK3 Cogne steel) is ammonia hardened, and the shafts are 35mm diameter. The "crankpin" is 30mm diameter.

On the driving side, the crankshaft is supported by a caged roller race 68mm in outside diameter, and on the timing side by a ball race of 80mm outside diameter. Both races are housed in bronze bushes pegged into the crankcase—as are the main gear bearings—to mitigate the difference in expansion rates between the steel of the bearings and the Elektron of the case. Circular balance weights are retained on the crankshaft by set screws.

The connecting rod is of chrome-nickel steel (CG4 Cogne steel) and is 175mm long between centres. The rod runs direct on uncaged steel rollers each 24mm long by 3mm in diameter, and is one of the most interesting features of the engine because the rod is of the split big-end type running direct on rollers—a system which is commonly regarded as impracticable.

However, handsome is as handsome does, and this split big-end design is a feature of all Guzzi four-stroke engines. Retaining bolts for the big-end cap are ground and are a precise fit in the reamered rod and cap holes. Before the bearing eye is ground, the cap is bolted up in position to a pre-determined pressure.

For a racing engine in the hands of a private rider, there is one very great advantage of this arrangement—he never has to do the one job which is usually quite outside the capabilities of any save the most skilled mechanics, i.e., align the crankshafts. Uncaged rollers are another unusual feature, but in practice the life of a big-end bearing is not abnormally short. On dismantling a used engine there is always evidence of "scuffing" of the rollers on the crankshaft journal, which makes one a little anxious at first. However, this is unimportant and good service is given before the shaft needs re-grinding.

Massive Flywheel

Primary drive is by spur gears, and the gear case carries an outrigger roller race to support the massive flywheel which is another unusual feature of all four-stroke Guzzis. It is perhaps of interest that while the makers are willing to disclose such closely guarded "secrets" as the size of the valves and the lift of the cams, they remain silent as to the weight of the flywheel! A left-hand threaded locking ring for the flywheel nut also serves as an extractor for the flywheel, which is keyed on the crankshaft taper. The flywheel drives the primary pinion through two dogs, which are kept in mesh by a spring at the back of the pinion. A felt washer in an annulus shrouding the shaft prevents all save a minimum of oil from seeping through from the crankcase to the gear case. As will be deduced from the primary-drive arrangement, the crankshaft runs "backwards."

Valve Timing

On the timing side, the crankshaft protrudes into the timing chest and carries a small key to locate the parallel-fitting half-time pinion, which is finished with three slots not quite equidistant to allow minute variations of valve timing. As the half-time pinion is reversible, six alternative positions are, in fact, available, and this suffices for an engine not over-sensitive to valve timing. A light, driving fit in the crankcase, the cam bearing pin has a small cutaway in its flange which locates on a dowel. The other end of the cam pin locates in a hardened bush pressed into the timing cover.

Uncaged rollers 35mm long by 3.5mm diameter are employed between the pin and the pinion with its two integral cams. Width of the cams is 9mm and both give a lift of 10mm. Diameter of the cam base circles is 45mm and the length of the cam-track is about 6¼ inches—possibly the longest in any racing motor cycle engine.

The cam-followers are fabricated out of

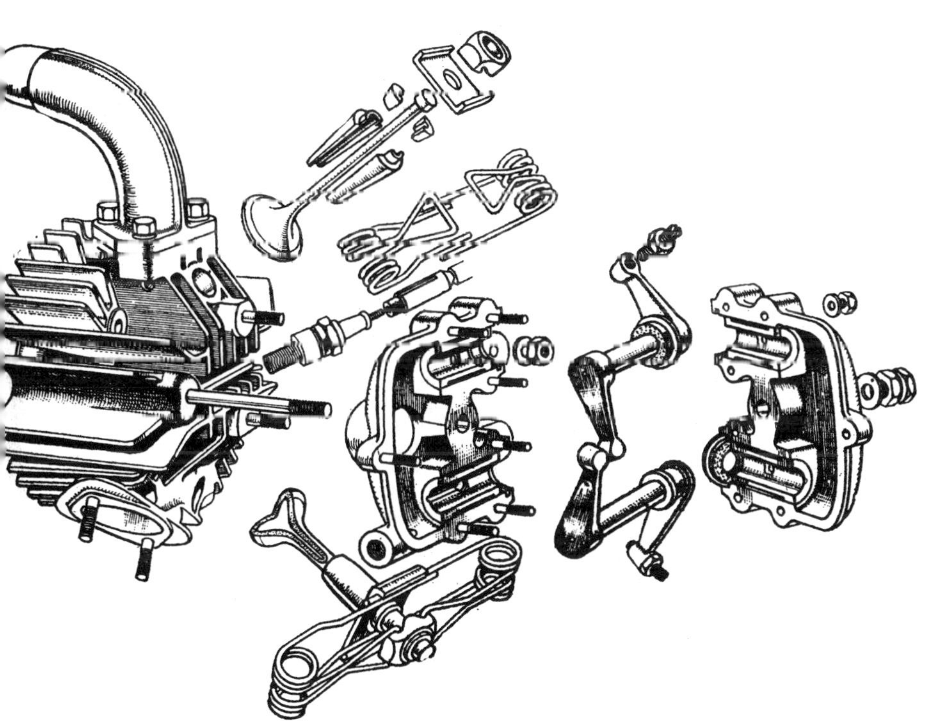

500 c.c. Gambalunga Guzzi . . . Continued

the solid. A 20mm-diameter, hardened-steel roller runs on each cam. The roller operates on a floating, bronze bush supported by a soft pin riveted in position. Push-rods are in chrome-molybdenum tube of 9mm diameter furnished with hardened-steel ball ends top and bottom. The cam pinion drives (above it) the magneto pinion. A light-alloy stirrup strap retains the magneto, and a felt washer at the back of the pinion prevents oil leakage from the timing chest. The magneto pinion retaining nut drives the revolution indicator gear box in conventional manner.

Situated below the half-time pinion, and driven by it, is the oil-pump pinion. The oil-pump is unusual because the feed side has gears, and the pickup side has spring-loaded "paddles." Oil is fed through the crankshaft and thence to the big-end bearing by way of a long jet. Oil escaping from the big-end lubricates the piston and cylinder walls, and the gear box pinions. The push-rods and rockers are lubricated by oil mist flung off the timing gears, which is returned by an external oil-pipe. Circulation of oil at racing speeds is about 13 gallons an hour.

Sodium-filled Valve

Cylinder and cylinder head are in aluminium-alloy (Idronalio 5 I), the former having a heat-treated nickel-cast-iron liner. The piston is of chill-moulded die-cast Lo-Ex alloy with the exceptionally low bare weight for such a large bore of 337 grammes (12oz). Two compression rings 1.5mm wide and one oil scraper 2.5mm wide are fitted. The piston is flat-topped with small cutaways for the valves, and the underside of the crown is lavishly ribbed. The gudgeon pin, which runs in a bronze bush pressed into the connecting-rod eye, has a diameter of 20mm and is located by circlips.

Internal form of the cylinder head is exceptionally shallow, and this, allied to the flat-top piston, gives a combustion chamber at the moment of ignition which one would have thought to have been far from ideal. The valves are set at an angle of 60 deg. to each other, and both valves have 120 deg. seats (included angle).

The inlet valve is an imposing affair of no less than 46mm diameter. It is made of chrome-silicon steel (S10X Cogne). The stem measures 8mm across, and runs up to a 10mm-diameter integral collar to support the collets; the collar is stellited. A split, bronze valve-guide is used. Measurements of the exhaust valve are 40mm diameter at the head and 11mm diameter at the stem, but in this case the valve is fined off to 9mm diameter to accommodate the collets under the stellited collar so that a normal one-piece guide can be used. A thin, bronze liner is fitted in the exhaust valve guide. The exhaust valve is sodium filled and made of chrome-nickel (D10X Cogne) steel.

The interesting feature is that the sodium cavity follows approximately the external shape of the valve, with resultant light weight and rapid heat flow, but how this desirable cavity is attained—well, one must guess! The crown is *not* welded to the head. Width of the valve seats is about 2.5mm.

Hairpin valve springs of 4mm shot-peened wire are used. At the collet end they are supported by aluminium collars, and the free legs lie in steel plates fitting round the valve guides and seating on the cylinder head. The installed pressure of the springs is 70 kgs (154lb) with the valves shut and 90 kgs (198lb) at full lift.

The long induction pipe, so typical of Guzzi machines, is lapped to the cylinder head and is a compound of steel tube and aluminium casting. Bore of the carburettor used on this engine is 32mm (1.26in) for the original Gambalunga and 35mm (1.38in) for the Tipo Faenza. Some of the engines have the induction pipe fitted by spigot into the head. To make the joint between cylinder and head, a copper gasket is employed and four long studs go right through from cylinder head to crankcase The rocker box is split, and carries forged-steel rockers in split bronze bushes. Felt washers retain the oil where the rockers emerge from the rocker box.

Push-rod Length

The joint between push-rod cover and rocker-box is made by a softish synthetic rubber washer to allow for variations in length due to changes of compression ratio. Rocker clearance is adjusted by 6mm diameter, fine-thread, hardened-steel screws (with lock nuts) which bear on the valve-stem tips.

A most irritating point about the Gambalunga engine is the need for constant adjustment of push-rod length and rocker clearance. In these days of widely varying qualities of fuel for different races, it is frequently necessary to change the compression ratio, and, of course, with each change of ratio there must be compensating change in push-rod length. Then the different co-efficients of expansion of the cylinder barrel, valves, and push-rods further make it necessary to adjust the rockers for varying engine temperatures.

Power Output

Clearances required for racing temperatures are about 0.002in for the inlet valve and about 0.012in for the exhaust valve; however, when the engine cools these clearances not only disappear completely, but both valves are lifted right off their seats. The result is that before the engine is started from cold the adjusting screws have to be slackened off, the engine must be warmed up, and the clearances reset. When one considers that a rider may be called upon to do this half-a-dozen times a day, when he is pressed for time getting gear ratios, jets, and so on, right for a race, it can be seen to be quite a major source of annoyance.

Before leaving the top of the engine (or, as it is a Guzzi, perhaps one should say the front) it may be apposite to say something about the power. On 73-octane fuel the compression ratio is 8 to 1, and both types of engine produce 34 b.h.p.

It should be noted that this is power at the final drive, not at the engine shaft. The Tipo Faenza produces this power at about 5,500 r.p.m., and the original Gambalunga at 5,800. In both cases the safe maximum is 6,000 r.p.m., though the engines will actually go climbing on far above this figure if allowed to do so, and on at least one occasion I have used "emergency revolutions" of 7,000 r.p.m. on the last lap of a race. The race was won—at the cost of a cracked crankcase!

The engine is one that responds to good fuel, and the use of 80-octane plus a higher ratio has a most marked effect on performance, particularly low down in the range. In common with other Guzzis, there is "tromboning" effect low down, even though a straight exhaust pipe is used, but this disappears at 2,000-2,500 revolutions and is rarely troublesome.

Layout of the gears is fairly conventional; and all that one can say as to gear ratios is that it is a question of "'orses for courses," as a wide range of both overall and intermediate ratios is available. Supported by a ball-race, the mainshaft protrudes through into the primary gear case, and the clutch back-plate is keyed to it. When the clutch is withdrawn, the pinion runs on a roller bearing.

On each side of the clutch pinion there is a Ferodo disc, and then alternate steel and copper plates—the copper plates are gashed to prevent them from "picking-up." The clutch withdrawal rod runs right through the mainshaft, where twin coil springs pull the clutch home. A tiny, caged ball-race on the end is shrouded by an end cap against which the clutch-operating lever pushes.

Many Successes

The layshaft of the gears runs in ball races, and, from the rear jointing faces of the crankcase, drillings run in behind these ball races. A certain amount of oil finds its way through the races and, by a channel, emerges just behind the final drive sprocket and so on to the chain.

Gambalunga machines have gained many successes in the field of "independent rider" racing. Such successes have been due to the racing efficiency of the machine as a whole and not, as is sometimes the case, to sheer torque at the end of the crankshaft. Without doubt, the greatest success by the Gambalunga was when Lorenzetti won the Ulster Grand Prix in 1948. A racing win is never a fluke, but this was something quite exceptional, for the Gambalunga has never been the "Big Gun" of the Guzzi factory. The twin - cylinder machines have always been able to see off the singles, which have done most of their campaigning in the field of "major-minor" racing.

During the past two seasons the singles have been practically unbeaten in such events. There can be no doubt that their success has been due not so much to very great power from the engine—though the power is good, especially for a push-rod unit, and the range of power is exceptionally good—but to making the best possible use of what power there is. The weight of the complete machine is only 253lb.

Moto-Guzzi 65

A Diminutive Italian Machine of Outstanding All-round Performance for Its Size

SMALLEST-capacity, fully-fledged, large-production motor cycle in the world, the Italian Moto-Guzzi 65 has a range of usefulness that belies its engine size. The combination of an efficient power unit, a three-speed gear box and light weight, make the machine a thoroughly practical little motor cycle, not only for runabout purposes, but also for far more ambitious journeys.

Though of small overall dimensions, the Guzzi provides a commanding, man-size riding position that would do credit to many larger machines. This results from a combination of 30-inch saddle height, relatively low footrests, and a handlebar shape which gives a natural grip angle. The machine handles extraordinarily well, steers to a hair's breadth and, in virtue of its low weight, can be "thrown about" in complete safety.

The parallel-ruler type front fork has an adequate range of movement and the rear-springing is soft enough to absorb even minor road irregularities. On the machine under test proprietary friction dampers were fitted to the rear suspension, though these were loosened off to make them ineffective. The need for dampers was never felt and throughout the performance range on a variety of road surfaces the standard of road holding and comfort was of a very high order.

For cold-engine starting two or three digs on the starter pedal were required; a first-kick start was usual when the engine was hot. Mild piston slap ceased when the engine had been running for a few minutes to warm up, and it then idled reliably and with no undue four-stroking; the pick-up on opening the throttle was brisk and without hesitation. There is only slight noise increase at wide throttle openings over the pleasant, musical exhaust whisper heard at idling speeds.

Though the clutch gave all the indications of freeing readily and fully, it was difficult to avoid a slight "scrunch" when engaging bottom gear with the machine stationary and the engine ticking-over. Operation of the clutch by means of a full-size handlebar lever was finger light and the clutch took up the drive easily and smoothly. The gear-change lever is situated on the right-hand side and just in front of the fuel tank, where it is well clear of the rider's knee, yet readily to hand. Changes could be made quickly and with little chance of missing the gear ratio required.

A very slight whine was audible in second gear at lowish speeds, but bottom gear (and, of course, the direct-drive top) and the helical primary drive from the engine to the gear box were noiseless. The brakes were outstandingly effective—light in operation, progressive in action, and capable of stopping the machine in almost unbelievably short distances; for example, in 22 feet from 25 m.p.h. In addition, when maximum braking was applied at high or low speeds the Guzzi remained completely stable.

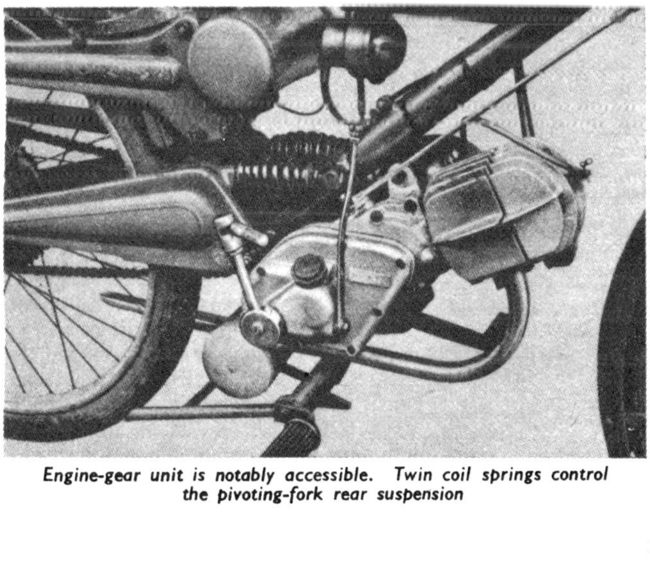

Engine-gear unit is notably accessible. Twin coil springs control the pivoting-fork rear suspension

The quiet liveliness of the engine, the good handling, the comfortable riding position, and the feeling of absolute command it engendered, combined to make the Guzzi a first-rate machine for traffic work. A speed of 25 m.p.h. from rest could be reached with a slickness comparable with average four-wheel vehicles, so that in built-up areas one could remain comfortably with the rest of the traffic.

On out-of-town roads an effortless cruising speed was between 25 and 30 m.p.h., with considerably higher speeds practicable in only slightly favourable conditions. For example, on two occasions with a barely perceptible tail wind on level roads speeds of over 35 m.p.h. were maintained for lengthy periods; although engine r.p.m. were well over 5,500, there was no trace of vibration nor any manifestation of over-driving.

INFORMATION PANEL

SPECIFICATION: Guzzi 65 c.c. (42 x 46 mm) two-stroke engine with three-speed gear in unit and helical-pinion primary drive. Detachable light-alloy cylinder head and light-alloy cylinder barrel with liner. Crankshaft mounted in phosphor-bronze and roller bearings; needle-roller big-end bearing. Flat-top light-alloy piston. Induction controlled by crankshaft valve. Gear-change by hand lever. Petroil lubrication, ratio 20 to 1. Tank capacity, 1½ gallons. Dellorto carburettor with lever air and throttle controls; air filter. Marelli flywheel magneto with lighting coils. Link-action front fork and pivoting rear suspension. Pirelli 26 x 1¾ x 2in tyres. Overall gear ratios 26.6, 17.4 and 10.2 to 1. Saddle height, 30in unladen. Wheelbase, 47in. Weight as tested with tools and 1 gallon fuel, 126lb.

PETROIL CONSUMPTION: Approximately 100 m.p.g. under hard riding conditions.

PRICE: £95, including import duty and Purchase Tax.

MAKERS: Moto-Guzzi, S.A., Mandello del Lario, Como, Italy. Concessionaires in Great Britain: Bob Foster, 472-4, Ashley Road, Parkstone, Dorset.

As might be expected, adverse winds and hills took their toll of the performance. In such circumstances one had the option of making good use of the gear box or letting the engine slog in top gear (assuming the gradient was not too severe), which it would do very happily and comparatively powerfully in the manner of two-strokes. A gradient that would defeat the Guzzi in bottom gear would rarely be encountered in hilly country given a modicum of forethought in route selection.

For experiment, the 1 in 5½ gradient of Pebblecombe Hill, Surrey, was tackled and surmounted with 18 unhurried "foots" while the engine pulled steadily in bottom gear. This was a good test because the hill is lengthy and, because bottom gear had to be engaged before the steepest part was encountered there was no possibility of the performance being falsified by rushing tactics. In full wet-weather riding gear, including waders, the rider's weight was 13-stone.

Fuel consumption under fairly hard riding conditions worked out at about 100 m.p.g. The fact is that the "big-machine" feel of the Guzzi, coupled with its many other attributes, tends to encourage full use of its performance—hence the "hard riding."

Finish of the 65 is in the well-known Guzzi red enamel, with chromium plate for the handlebar and controls, exhaust pipe, wheel rims and lamp rims. There is a useful parcels grid on the tank-top and a sturdy carrier on the rear mudguard.

The 98 c.c. Two-stroke

GUZZI "ZIGOLO"

"Motor Cycling" Records Impressions of a Quality Italian Ultra-lightweight with Ultra-modern Features

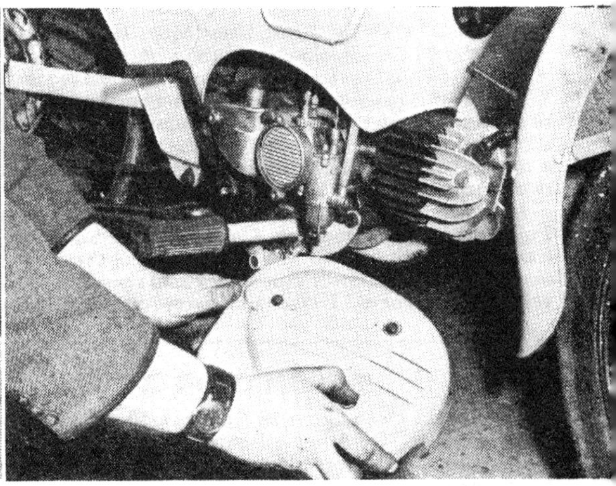

The Guzzi "Zigolo" (above) provides clean, comfortable transport for the everyday rider, combined with a performance which would commend itself to the lightweight motorcycle enthusiast. (Above right) Removal of the right-hand shield discloses the Dell'Orto carburetter, with integral air filter and sludge trap. Above the shield is the door giving access to the fuel tap.

TESTER'S ROAD REPORT

Maximum Speeds in:— (Time from Standing Start)

Top Gear (Ratio 7·83 to 1) 45 m.p.h. = 4,870 r.p.m. 50 secs.

Second Gear (Ratio 13·42 to 1) 34 m.p.h. = 6,300 r.p.m. 34 secs.

Speeds over measured Quarter Mile:—

Flying Start 43·6 m.p.h. Standing Start 29·5 m.p.h.

Braking Figures On DRY TARMACADAM, Surface, from 30 m.p.h.:—

Both Brakes 36 ft. Front Brake 56 ft. Rear Brake 58 ft.

Fuel Consumption:—

30 m.p.h.—160 m.p.g. 40 m.p.h.—112 m.p.g. 50 m.p.h. —— m.p.g.

BRIEF SPECIFICATION

Engine: Guzzi air-cooled single-cylinder two-stroke, horizontally mounted; bore 50 mm. by 50 mm. stroke = 98 c.c.; C.R. 6 : 1; cast-iron barrel; light-alloy cylinder head; deflectorless light-alloy piston; rotary valve for mixture induction; petroil lubrication; Dell'Orto carburetter.

Gearbox: Guzzi three-speed foot-controlled positive stop gearbox built in unit with engine; gear ratios 7.83, 13.42 and 20.57 to 1; multi-plate clutch, running in oil, driven by helical gears from crankshaft; final drive by chain.

Ignition and Lighting: From Marelli flywheel magneto-generator, equipped with lighting coils and separate ignition coil; Aprilia headlamp; tail lamp; lights and horn controlled by handlebar switch.

Frame: Of beam type, with swinging-fork rear suspension controlled for shock and rebound by central rubber springs; friction dampers at rear; sub-frame formed by steel pressing, incorporating under-tank shielding, with built-in tool box and "cupboard" for fuel tap (three-position) and extended float chamber tickler; front forks Guzzi telescopic type.

Wheels and brakes: Chromium-plated 19-in. rims front and rear, carrying 2.50-in. by 19-in. Pirelli tyres; internal-expanding brakes, 5-in. diameter.

Tank: Of approx. 3 gal. capacity.

Finish: Forks, mudguards, tank, handlebars, frame pressings in light grey enamel, with Moto Guzzi motif on tank sides and on front and rear mudguards; wheel rims chromium plated, brake back-plates in polished light alloy; headlamp black.

Dimensions: Wheelbase, 50 in.; overall length, 77 in.; saddle height, 30 in.; overall width, 26¾ in.; ground clearance, 6½ in.; weight, 172 lb.

Manufacturers: Moto Guzzi, s.p.a., Mandello del Lario, Como, Italy.

Concessionaire: Bob Foster, 472-4 Ashley Road, Parkstone, Dorset.

Price: £110, including Purchase Tax.

Equipment: Electric horn, built into under-tank shielding; tool kit.

Annual Tax: 4s. 10d. (per quarter).

Extra: Smiths 55 m.p.h. speedometer, £4 10s.

BIGGEST of all the Italian motorcycle concerns is the s.p.a. Moto Guzzi, whose annual production ranks third only to the two scooter giants—Innocenti and Piaggio. Since 1921, when the marque was introduced, Guzzi designers have concentrated upon the horizontal engine layout which has, over the years, come to be regarded almost as a Guzzi hall-mark. Not surprisingly, therefore, their latest product—the 98 c.c. two-stroke "Zigolo," introduced last year—follows their now well-established style.

Although the extensive use of steel shielding conceals it from immediate observation, the "Zigolo" has, in fact, a beam-type frame equipped with swinging-fork rear suspension. The "sub-frame," which carries the rider's saddle and the rear carrier, also forms the rear mudguard and is extended forward to become a full under-tank shield, with an in-built tool box on the left and—on the right-hand side—a door giving access to the fuel tap and extended float-chamber tickler. The movement of the swinging-fork (a member fabricated from oval-section steel tubing) is controlled by a large-diameter rubber spring working in compression and contained within the "bodywork" close to the pivot point.

At the front, suspension is by telescopic forks, of quite conventional design except for the "muffs" to which the wheel spindle is attached. These are of light alloy and each is held, by one nut, to a keyed and threaded extension of the lower slider. Thus the removal of two

operated rocking-gear ch
clutch take-up perhaps a
tion of both controls w
Once under way, it wa
rapidly through the ge
happily in top at so
20 m.p.h., the flexible
pulling well at such a g
respectable degree of ac
attained by holding on t
the unit would rev. fre
the highest gear being
" peak " had almost beer
With these dual attrib
speed pulling and an er
wisely-regulated design l
of a praiseworthy degree
" Galletto " proved to be
use in traffic. Such gaps a
the clanking Milanese
bicycles, scooters, Fiats
wheelers which spread the
po fido roads could imr
advantage of . . . and, a
the scale, when conditio
there was no need to jug
Coupled with this plea
a degree of handleability
compensated for the unf
of the road " driving.
few utility models which
this respect as does the
is delightful—even thou
frame, one is robbed of k
be cornered like a spo
until the footboards we
the deck, and thrown fi
the true good-steering s
from a leading racing f
Aiding this good-mann
the first-class suspension,
ment sufficiently long to

Engine: 192 c.c. Guzz
65 mm. by stroke
cylinder and head
push-rod operated;
b.h.p. 7.5/5.200 r.p.
buretter; instrument
filter and hidden be
Transmission: Gearbox i
positive-stop footc
pedal; primary drive
by ½-in. by .305-in.
Frame: Welded up from
steel channel section
work " of mingled s
Lubrication: Dry-sump
mechanical oil pump

nuts, and the detaching of the front brake cable, is all that is required to effect the removal of the front wheel.

As already mentioned, the engine-gearbox unit is horizontally mounted, with the single air-cooled cylinder protruding forward of the shields. It comprises a single-port two-stroke engine, with " square " bore and stroke dimensions of 50 mm. giving a capacity of 98 c.c. An interesting feature is the use of rotary valve induction, the valve being incorporated in the engine shaft on the right-hand side. The Dell'Orto carburetter fits directly on to the crankcase and is partially enclosed by a shield.

Ratios of 7.83, 13.42 and 20.57 to 1 are supplied by the gearbox, which is of positive-stop, foot-controlled type.

Current for ignition and lighting is supplied by a Marelli flywheel magneto-generator driven direct from the crankshaft on the left-hand side. The ignition coil is remotely mounted and is reached through the previously-mentioned door in the right-hand shields. Like the carburetter, the " electrics " are protected by a readily detachable pressed-steel shield.

An Aprilia headlamp of approx. 5½ in. diameter is employed. This has provision for the incorporation of a speedometer within the headlamp shell, but on the model tested a Smiths 55 m.p.h. speedometer, with external gear drive, was fitted.

Lighting and the electric horn—which component is neatly built into the under-tank shielding—are both controlled from a neat " cluster " fitting on the left handlebar. Clutch, throttle and front brake controls are all inbuilt and, in the case of lever controls, the cable adjusters are mounted on the handlebar static member.

Trouble-free

In the hands of *Motor Cycling's* testers, the " Zigolo " covered close on 2,000 miles during the six weeks in which it was " on the strength " and throughout that time—with the sole exception of an inexplicable bout of plug-whiskering which recurred on isolated occasions—it provided trouble-free service.

Starting from cold proved to be quite easy—the carburetter was generously flooded, the air lever closed, and two dabs on the kick-starter would bring the engine to life. It proved advisable to allow time for the unit to warm up before opening the air lever.

First-gear engagement was at all times positive (the operation was on the " down for down, up for up " principle) and the change into second gear could be made as rapidly as the rider wished. Upward and downward changes between second and third gears, however, demanded more care, and it was advisable to be slow and deliberate if missing the gear was to be avoided. The clutch operation proved to be of the type to which that over-worked adjective " silky " could justly be applied.

Smart, ultra-modern lines are outstanding features of the latest model from the Guzzi stables.

Not once throughout the test did the operating cable require adjustment.

Smooth and flexible, the little engine provided an adequate degree of acceleration in the intermediate gears. Top gear could be held down to some 15 m.p.h., but generally speaking it was found advantageous to engage second gear for speeds below 20 m.p.h. Trickling with the clutch fully home could be indulged in with first gear engaged and the " Zigolo " moving at a fair walking pace.

Part of the machine's duties while on test consisted of providing daily transport for a staff man resident near Brighton—a trip of some 60 miles morning and night. The little Italian lightweight withstood the task manfully, recording a figure of 1 hr. 50 min. door-to-door with unfailing regularity, no matter what the weather conditions. On the open road part of the run, a 36-mile stretch was regularly reeled off in exactly one hour, despite the need for traversing at least three long, stiff climbs.

Here, the model's pulling power came into play. It was found capable of tackling quite sizeable gradients in top gear, its ability to pull steadily at around the 20 m.p.h. mark being particularly noted. Second gear, unfortunately, often proved too low for the job in hand, and the tester sighed more than once for a fourth ratio mid-way between top and second gears. On the run mentioned, the model was generally cruised at between 40 and 45 m.p.h. as prevailing conditions permitted, though on downhill stretches it was not uncommon for the speedometer needle to go " off the clock." Overall fuel consumption on these journeys was in the " over 100-m.p.g. " class. Except at peak revs. in each gear, the unit was nearly vibrationless.

Handling, whether on dry roads, on wet cobbles, or latterly on iced and snow-covered highways, was beyond reproach. The " Zigolo " steered perfectly, held its line to a hair, and never did more than step sedately out by two or three inches at the rear when cornering on really slippery surfaces. Much credit for this must go to the suspension systems, which absorbed all but the worst in the way of surface inequalities.

Good Lights

Night-time riding was made a pleasure by what can only be described as a superlative lighting set. Indeed, for the first time the tester really felt on terms of equality with car drivers in the matter of night vision, for the Aprilia headlamp threw a powerful and penetrating beam which gave almost perfect illumination of the road for well over 100 yards ahead, while for town driving even the dipped beam was an embarrassment, bringing a volley of flashes from approaching traffic even though the lamp had been tilted to point only a few yards ahead!

Apart from broken filaments, which entailed the replacement of no fewer than four tail bulbs, the electrical equipment was trouble-free throughout the test.

Considering its extensive nature, the mudguarding was frankly disappointing. Spray, thrown forward by the front wheel and caught by the airflow, tended to blow back clear of the front mudguard.

No means of adjustment is provided for the riding position, the saddle, foot rests and controls all being mounted in a set relationship. While this " in-built " position is quite comfortable for a rider in the " short-to-medium " height range, it was felt that a taller man might experience some discomfort.

Summing up, the Guzzi " Zigolo " is an ultra-modern ultra-lightweight, with design characteristics which place it high in the luxury touring class, a performance which belies its small engine capacity, and a considerable degree of economy to add to its appeal.

IMPRESSIONS (

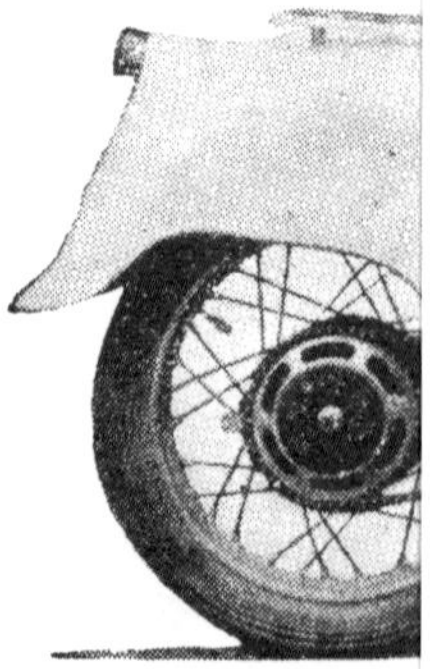

PROBABLY the mo
design to appear i
production version of tl
—then a three-speed ove
150 c.c. It represente
"marriage" of the t
motorcycle and scooter
which would have been
not the traditional Gu
layout lent itself so admi
Wheel sizes, though s
the average lightweigh
larger than those of t
Motorcycle-type botto
were employed, but
utilized a swinging a
carried on a stub ax
horizontal engine, with
gearbox, the calls of wei
be adequately met, a
important open-frame l
Cooling of the cyli
problems, for the engine
excellent air conduit.
It is not surprising
eminently practical poir
"Galletto" soon beca
Subsequent development
configuration unaltered
simply at augmenting
especially acceleration a
Thus the capacity
to 160 c.c.; then to 1
192 c.c., the latest engir
of 65-mm. bore by 58-m
with the 60-mm. by 53-
the original model, havi
by stages of 62 mm. by 5
65 mm. by 53 mm. (175.
a fourth ratio has been
That these latest mc
fact, been beneficial w
points to strike the test
chance, had been respons
figures obtained in th
Cycling's test of the 160
in our issue of August
collected the new mod
depot of Moto Guzzi s.
November last.
Immediately noticeabl
the exhaust note, thou
pleasant crispness, was
that mechanical noise
earlier type) had been b
First-gear selection—b

CYCLE

ROAD TEST NO. 107

By the CYCLE Staff

After the 4,000 mile test, Stan Barr treated the Moto-Guzzi like an old friend. Here he demonstrates the bike's agility around a typical California street corner.

MOTO GUZZI

LODOLA 235cc GRAN TURISMO

Quiet even to the most delicate ears. The exhaust note is held to a minimum by the ultra-long muffler.

PERHAPS one of the most extensive road tests we will ever write was recently conducted on a Moto Guzzi Lodola 235cc Gran Turismo—ridden cross country by Stan Barr (weight 170 lbs.), via New Orleans, for a total of 3,947 miles which included some side trips from New York City to Los Angeles.

The weather that day we last saw the skyscrapers was sunny, but cool gusts of late winter wind held the thermometer at a steady 50 degrees. Not at all bad weather for cycling, if you're dressed properly. A rather uncommon, but very practical Barbour Suit was worn, insuring protection against cold, wind, and rain.

Our test bike's speedometer read only 740 miles as we picked up the entrance ticket to the New Jersey Turnpike. The Turnpike, for readers unfamiliar with major Eastern roads, is a mostly four lane monontonous monster that diagonally slashes Jersey from North-East to South-West. For almost 2½ continuous hours, the speedo needle was held between 50 and 55 mph, as cars, trucks and buses buzzed past, near or over the 60 mile limit. During this stretch, a rather disappointing top end of 65 was briefly recorded. Considering the newness of the Moto Guzzi, however, we had hopes of reaching a better top speed as more and more miles were covered.

Handlebars are slightly raised and were found to be very comfortable for touring. Available on special order are low bars, which come as standard equipment on the Lodola Sport, virtually the same motorcycle but with a 175cc overhead cam engine. Right and left hand control levers are in the conventional positions but shifting is performed by the right foot operating a rocker-type shifting arm. The pattern is down with the heel for first, and down with the toe for second, third and fourth. Neutral sits between first and second. Until we got used to this system we were able to double check when we thought we were in neutral by the verticle position of an indicator. Downshifting is a snap as three clicks with the heel bring you back to first.

In Bethesda, Maryland, just outside of the National Capital we stopped at Guzzi dealer Ed Miller who gave the bike its first 1,000 mile oil change and adjusted the drive chain. We headed now for the beginning of the Blue Ridge Skyway, to test the 235cc's ability to climb hills and its agility on winding roads. Although the Skyway was completely void of traffic in either direction, a light rain mixed with heavy fog on the westerly side of the ridge prevented any evaluation of high speed cornering. Acceleration for a 235cc bike, up the steep grades from a standstill was very quick and positive with sudden lurches of power in the first three gears. Fourth, as we expected, was sluggish.

On one occasion, a young motorcycle buff remarked after studying the single cylinder engine inclined at 45 degrees, "Man, someone put your engine in crooked." We rather like the way the engine slopes and feel that it adds a racy quality to the graceful lines of the cradle-type frame and sculptured gas tank. The Guzzi people have installed an oversquare mill, with 64mm stroke and 68mm bore. It produces a rather low 11 horses at 7.5 C.R. With low rpm and with the old reliable push rod valve operation it is a safe guess that this engine will run for a long, long time.

The rain continued even when we left the Blue Ridge Skyway to cut over

Editor Stan Barr applies the excellent stoppers, bringing the Moto-Guzzi to a positive halt. Note dipping of the front fender which is attached to the forks.

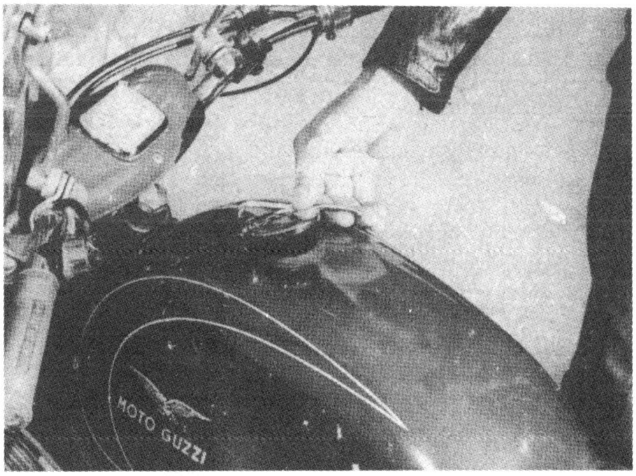

Opening to the 3½ gallon gas tank is by spring-loaded filler cap. Horn button and high and low headlight controls are thumb operated on the left handlebar.

to route 11; even so, the Lodola purred smoothly, without a miss—an excellent weatherproof ignition system. One night just past Knoxville, Tenn., we had our first and only mechanical breakdown when a clutch cable snapped. Sitting back in New York was a forgotten spare cable so, reluctantly, we called an SOS to the nearest dealer, Herb Hitch, back in Knoxville. Herb left his T.V. and was waiting at his shop when we arrived. He made a make-shift cable which as far as we know is still holding together. (This little incident typifies the fraternity-like feeling between most dealers and riders).

For six days in all we were plagued by the rain, and it only let up on the outskirts of the City of New Orleans. During these cold and wet days, little or no effort was required to light a fire in the cold engine. We were amazed how time and time again the cold engine started with just the first kick.

Going through Texas the top speed of the Guzzi was gradually increasing. Long runs of 70 miles per hour were common although one State Trooper confirmed (as a favor) that the actual speed was 68 mph. There is a modern light alloy head over the cast-iron block and no sign of compression loss was noticed where the two different metals joined, even though we were now pushing the bike rather hard.

Once or twice we were forced to give the Guzzi an unexpected braking test as our path was suddenly blocked by unaware motorists. Large full width hubs fore and aft hold plenty of stopping stuff. The hubs are finned for cooling and although the brakes give out with a funny squeal on occasions, due to the lining material, we found them to be excellent and they never faded once during the close to 4,000 miles covered.

A flat rear tire in the middle of the Western Texas wastelands delayed us for exactly 30 minutes. The rear wheel is easily removed by slipping the axle out and pulling the wheel off six rubber covered studs. Tire changing tools and a hand pump, which came with the not too adequate tool kit, must have been tested thoroughly for they worked like a charm. This operation can naturally be accomplished faster, but we were not out to set a record.

In El Paso, Texas we pulled into Stark's Cycle Center for another oil change which was well overdue. Here we also switched the 25 watt headlight bulb for a 35 watt. The battery-generator ignition produces enough juice to take the larger bulb and this is a recommended procedure for all Guzzi owners.

Across Arizona with the bike now

"This Italian lightweight is an ideal motorcycle for women," beamed Carol, as she returned from her first ride on the Moto-Guzzi Gran Turismo.

Easy starting is demonstrated by gal editor, Carol Anderson. Pannier racks, windshield, exposed rear shocks are the only differences from a stock bike.

Moto Guzzi seat hinges up to reveal tire pump; access to oil filler cap is also under seat. Rocker-type center stand is well-designed.

A Barbour Suit insured protection against cold, wind and rain during the New York City to Los Angeles test.

fully broken-in we were prepared to make some final speed and acceleration tests. Digging out on a flat straightaway, with a five mile per hour headwind, we were pleasantly surprised at how quickly the engine now wound up in all four gears. Popping the clutch while in first sent the Guzzi up to 30 mph. Number two gear took it up to 44 mph, third to 61 mph and a top of 80 mph in fourth. There is fantastic low-end torque and the 235 Guzzi buzzes up to 60 mph in 14 seconds—excellent for its capacity.

Fuel consumption at 60 miles per hour was 62 miles per gallon. Gas mileage increased to an economical 85 miles per gallon at 40 mph. Primary drive is by costly helical gears eliminating the adjustment problems of primary chain drive cycles. Cornering is fast and the Guzzi handles so well that we once scraped the muffler slightly trying to find the limit of adhesion. The ultra-long muffler, with a pinky-size hole at the end gives out with a quiet exhaust note. A leverage principle has been incorporated in the center stand, so that by shifting your weight on one part of the stand, the bike will rock back into place. On the demerit side, we found that the self-oiling device was inadequate in keeping the chain well lubricated.

Summing up—The 235cc Gran Turismo Moto Guzzi is a well thought out piece of Italian machinery that lives up to its world-wide reputation of being a quality motorcycle.

Our test bike was supplied in New York by the Berti Corporation, sole importers of Moto Guzzi, located at 124 Covert Avenue in New Hyde Park.

Close-up of the 235cc engine also shows cradle-type frame, light alloy cylinder head and Dellorto carburetor.

MOTO GUZZI 235cc GRAN TURISMO SPECIFICATIONS

ENGINE: 235cc four-stroke, single cylinder, inclined at 45 degrees. OHV with push rod operation. Compression Ratio 7.5 to 1. Bore 68mm. Stroke 64mm. Maximum B.H.P. about 11. Light alloy cylinder head.

IGNITION: Coil ignition fed either by the battery or generator.

TRANSMISSION: 4 speed gear box, foot operated. Constant mesh gears with front dogs.

SUSPENSION: Telescopic front fork and rear swinging arm with hydraulic dampers, adjustable for two positions.

CLUTCH: Multiplate clutch running in oil bath.

CARBURETOR: Connected to an inlet silencing chamber incorporating an air filter.

BRAKES: Expanding type. Full width hubs, with cooling fins.

CAPACITIES: Gas Tank—3 1/2 gallons.

Oil Tank: 3/4 gallon.

WEIGHT: 253 lbs.

FUEL CONSUMPTION: 85 m.p.g. (average)

PERFORMANCE: Maximum safe speed in gears:

1st 30 mph
2nd 44 mph
3rd 61 mph
4th 80 mph

NATIONAL DISTRIBUTOR: The Berti Corporation, 124 Covert Avenue, New Hyde Park, New York

ROAD TEST

DASHING ITALIAN

THE Moto Guzzi Zigolo is a dashing little Italian machine with a host of interesting technical features. The 110 c.c. two-stroke engine has its single cylinder pointing forward, and the carburetter is connected straight to the crankcase. There is no primary chain—drive is by gears to the gearbox, and chain from there to the rear wheel.

This all adds up to a neat, sporty-looking machine with a very low centre of gravity. The first thing you notice is the ease of starting. This is one result of the rotary valve induction system. The other is that as the petroil mixture is drawn straight into the crankcase you don't have to use as much oil as on conventional two-strokes. The recommended ratio is 50:1, giving better fuel consumption and fewer oiling troubles. We managed to get an overall consumption figure of 105 m.p.g., driven hard most of the time.

The Zigolo looks nice—bright red, beautifully finished with a massive 3 gallon fuel tank. Full marks too, for the quick release petrol cap.

On the road this Guzzi really handles. It is light, of course, and she steers to a hair both in traffic and at speed. The engine is very quiet—unusual for an Italian. This is due to a large extent to the inlet silencer fitted to the carburetter.

The riding position is comfortable, the "straight" bars giving good control without the discomforts of having to adopt a stooping crouch. Space for a pillion passenger is very limited, however.

There is a three-speed gearbox—and I must admit that I would have preferred a four-speed. Bottom and middle are reasonably close together, but top is very high indeed, and there is a big gap between middle and top. The engine has plenty of pulling power, mind you, and slogs happily at low revs. Given time the speed can be built up to a shade over 50 m.p.h.. and will stay there as long as conditions permit. And this is only a 110 c.c. engine, remember.

The horn makes a most peculiar noise—but it seems to attract attention, which is the main thing. The two little lids in the engine cowling come off easily, and reveal the tool box and fuel tap. There is no difficulty in lifting the Zigolo on to the centre stand provided you use the seat strap—the underside of the rear mudguard is as sharp as a knife!

Apart from the rocking gearchange pedal and left hand (or is it foot?) kick-starter the controls are quite conventional. One interesting point is that the engine can be started with the kickstart whilst still in gear—just pull the clutch out. Very useful when stalled at traffic lights.

To sum up—a small machine, cheap to buy and cheap to run, with an exciting performance for so small an engine.—
D. WEIGHTMAN.

HOW LONG DOES IT TAKE

Remove and replace head	8 minutes
Remove and replace rear wheel	7 minutes
Adjust both brakes	¼ minute
Check points	5 minutes
Adjust rear chain	2 minutes

'M.M.' MACHINE TEST REPORT No. 24

Machine Moto Guzzi Model Zigolo c.c. 110

Test Mileage 352 Price new £120 Examiner D. Weightman

Supplied by Motor Imports Ltd., 7 Gresham Road, London S.W.9

Maximum Points 10 (compared with machines in same c.c. and price range)

Brakes (front)	8	Lights	8
Brakes (rear)	9	Engine accessibility	7
Brakes (both)	8	General performance	9
Steering at high speed	9	Overall finish	9
Steering at low speed	9	Electrical layout	8
Gearbox action	7	TOTAL	91

Fuel consumption over whole test 105 m.p.g.

Acceleration 0-40 11½ secs.

Top speed 51 m.p.h.

Remarks

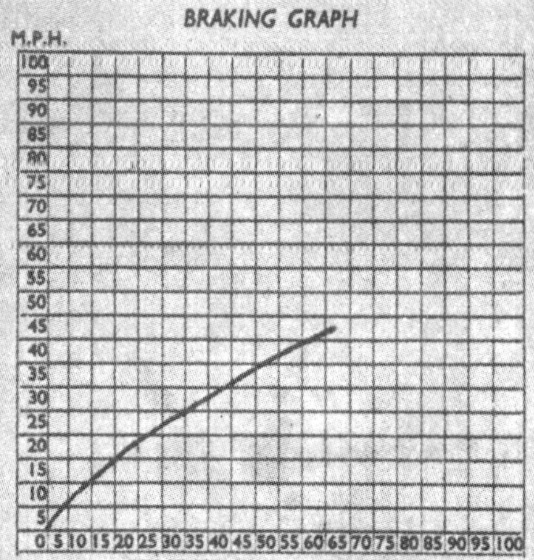

ROAD TESTS OF NEW MODELS

110 c.c. Moto-Guzzi Zigolo

Sprightly Italian Lightweight with Clean Lines, Excellent Roadholding and Powerful Brakes

YES, you *can* have clean, flowing lines without either increasing the weight or decreasing the accessibility! If you want proof, take a look at the Moto-Guzzi Zigolo 110. Study the way in which the two-stroke engine—mounted horizontally, cylinder head projecting forward—blends into the pressed-steel, body-and-mudguard unit, and how the body pressing is itself allied with the large-diameter tubular backbone.

But not only the appearance is interesting, for the 110 c.c. engine has an overhung crank, and a rotary inlet valve is incorporated on the right of the crankcase. The cylinder and head, both in light alloy, have deep, radial finning. The cylinder bore is hard-chromium plated. And a feature which shows up very favourably is the low petrol-oil ratio of 50 to 1.

Ultra-light the Zigolo may be, but clever design results in a riding position suitable for both long- and short-leg riders, while the relationship of handlebar, seat and footrests affords a relaxed stance with just the right amount of forward lean. The rocking-pedal gear lever can be operated readily by pivoting the foot on the right rest. However, the brake pedal has no adjustment and is set a little too high for comfort, while the clutch and front-brake levers—with pivot blocks welded to the handlebar and provided with finger cable adjusters—require a long reach. Grouped in a neat unit, the light switch, dipswitch and horn button are within easy reach of the left thumb.

Left: The light-alloy cover for the right-hand side of the crankcase serves also as the air-filter outer cover. A foam-plastic strip seals the gap between the cover and backplate—air is drawn through the tube at the top. Below: Although the footrests may appear to be too far forward the riding position proved comfortable

Access to the fuel tap is through a trapdoor in the right of the body (similarly, tools are carried in a small compartment on the left) and starting was just a matter of turning on the petroil; there was no need to flood the carburettor. The engine would fire on the first or second kick, even after a night in the open.

An endearing feature of the Moto-Guzzi unit was the way in which it would two-stroke happily whether under load or running light. Pick-up from low speeds was exceptionally good and the machine would accelerate smoothly and most willingly from 16 m.p.h. in top gear.

Long-distance journeys indicated that the dual-seat was not so comfortable as it might have been. As to the engine, best results were obtained by allowing it to buzz in the gears; used that way, there was ample power in hand, even with a ten-stone passenger added, to keep pace

with traffic. Solo, a cruising speed of just under 50 m.p.h. could be held without effort. Two-up, there was no difficulty in maintaining 40 to 45 m.p.h. Because of the wide spread of power, the three-speed gear box was adequate, and a well-chosen second ratio proved capable of coping with the steepest hills.

Movement of the gear lever was pleasantly short, but a trifle heavy, and for a noiseless change up it was advisable to let the engine revs die down before moving the pedal; clashing could otherwise occur. Light and smooth, the clutch seemed indestructible. Silky smooth, too, was the transmission, allowing the engine to pull sweetly at all speeds, with the only vibration a slight tremor at the handlebar at about 50 m.p.h. in top.

As might be expected where a horizontal engine layout (and thus a low centre of gravity) is adopted, roadholding and handling of the Zigolo were right off the top shelf. Sufficiently soft to absorb average road irregularities at low speeds, the front and rear springing adequately ironed out bumps at higher speeds without any tendency to pitch or bottom. In this respect, the hydraulic damping of the rear is a considerable improvement over the friction damping of the earlier 98 c.c. Zigolo. Only by hard braking over downhill bumps could the front fork be bottomed. Towards the end of the test a weep of oil appeared on the right fork leg.

Used together, the powerful brakes were more than adequate for the model's performance. The front brake was particularly impressive—light in operation, with a progressive bite and easily controllable on slippery going. The rear brake, however, tended to lock the wheel rather easily and called for a sensitive touch on the pedal. Neither brake lost any of its efficiency in heavy rain.

Lighting was a disappointment. The intensity and spread of the beam limited safe after-dark speeds to below 40 m.p.h. No provision is made for illuminating the speedometer. Cut-off from the dipped filament ensured that oncoming drivers were not dazzled, but there was an annoying time lag in dipswitch operation. The performance of the horn was inadequate for modern traffic conditions.

In a total test distance of over 600 miles, the rear chain was adjusted once—and that was all the maintenance required. But it was evident that an owner could carry out work on his machine with the minimum of trouble. For instance, the crankcase side covers are each retained by only three screws, and on removing the covers there is immediate access to the carburettor and gear box oil-filler plug. Contact-breaker points are reached by detaching an aluminium disc on the flywheel.

To clean the model calls for but two minutes' work with a damp cloth; and in any case, so effective is the mudguarding that very little road filth reaches the bodywork.

There, then, is the Zigolo 110, an inexpensive, workaday lightweight but one which has a charm of its own. It demonstrates that a utility mount can still have plenty of style and well deserves a place among today's outstanding ultra-lightweights.

Quietness is exemplary; note the large expansion box and small tail pipe

◀ *The gear-box oil level is easily checked on the dipstick. Removing an aluminium disc on the flywheel gives access to the contact-breaker points*

SPECIFICATION

ENGINE: Moto-Guzzi 110 c.c. (52 x 52mm) single-cylinder two-stroke with rotary inlet valve. Light-alloy cylinder head; light-alloy cylinder barrel with chromium-plated bore. Overhung crankshaft supported in one ball bearing; plain big-end bearing. Compression ratio, 7.5 to 1. Petroil lubrication; ratio, 50 to 1.

CARBURETTOR: Dellorto, with air filter. Air slide operated by handlebar lever

IGNITION and LIGHTING: Filso flywheel generator with lighting coils and remote-mounted ignition coil. CEV 5in-diameter headlamp with pre-focus light unit and 25/25-watt main bulb.

TRANSMISSION: Moto-Guzzi three-speed gear box in unit with engine; rocking-pedal, positive-stop control. Gear ratios: bottom, 19.4 to 1; second, 11.4 to 1; top, 7.38 to 1. Multi-plate wet clutch. Helical-gear primary drive; final drive by ½ x 0.305in chain with guard over top run. Engine r.p.m. at 30 m.p.h. in top gear, 3,300.

FUEL CAPACITY: 2½ gallons.

TYRES: Pirelli: 2.50 x 17in ribbed front, 2.75 x 17in studded rear.

BRAKES: 5½in-diameter (approx.) front and rear

SUSPENSION: Hydraulically damped, telescopic front fork; pivoted rear fork controlled by hydraulically damped spring units.

WHEELBASE: 49½in, unladen; ground clearance, 6in unladen.

SEAT: Prestigio dual-seat; unladen height, 30½in.

WEIGHT: 136 lb, fully equipped and with half a gallon of petroil.

PRICE: £99 9s 3d; with purchase tax (in Great Britain only), £119 19s 6d.

ROAD TAX: £1 a year.

CONCESSIONAIRES: Motor Imports Co., Ltd., 7, Gresham Road, London, S.W.9.

PERFORMANCE DATA

MEAN MAXIMUM SPEED: Bottom: 20 m.p.h.
Second: 34 m.p.h.
Top: 52 m.p.h.

HIGHEST ONE-WAY SPEED: 54 m.p.h. (conditions: gusty side wind; 13½ stone rider wearing two-piece suit and overboots).

MEAN ACCELERATION:

	10-20 m.p.h.	20-30 m.p.h.	30-40 m.p.h.
Bottom	3.4 sec	—	—
Second	4.2 sec	5.6 sec	—
Top	—	8.4 sec	8.8 sec

Mean speed at end of quarter mile from rest: 48 m.p.h.
Mean time to cover standing quarter mile: 26.2 sec.
Mean time taken from 0-30 m.p.h.: 7.4 sec.

PETROIL CONSUMPTION: At 30 m.p.h., 102 m.p.g.; at 40 m.p.h., 84 m.p.g.

BRAKING: From 30 m.p.h. to rest, 34 feet (surface, dry tarmac).

TURNING CIRCLE: 12ft.

MINIMUM NON-SNATCH SPEED: 11 m.p.h. in top gear.

WEIGHT PER C.C 1.24 lb.

A TRUE ITALIAN

George Wilson Samples an Trial 235 c.c. Moto-Guzzi

FUNNY, I thought, could I have lost count? I'd given the gear pedal an additional prod to make sure that top was in correct engagement and I'd found *another gear.* It was then that the old memory strings twanged. Hadn't I heard down in Wales during "International" week that the Moto-Guzzi Lodolas were five-speeders? So it was. Five cogs the little gem I was riding had. There I was, over in Milan, supposed to be making the rounds of the stands at the *Esposizione del Cicli e Motociclo.* But that Friday morning there wasn't much to see anyway. As usual with shows it seemed certain that the thing couldn't possibly be ready for the official opening the next morning. So I'd trundled out along the *autostrada* to Lecco, up that beautiful, winding road bordering Lake Como to Mandello del Lario—to the village that owes its very existence to the vast Moto-Guzzi factory. I'd telephoned the previous evening. And there, waiting for me, were an I.S.D.T. Lodola and a Stornello Sport 125, just asking to be ridden.

Here was an example of the enthusiasm with which a visitor is greeted in Italy, and particularly in Mandello. The I.S.D.T. models had all been stripped for inspection. But between the time of my call on the Thursday night and when I arrived at the factory, an engine had been reassembled and fitted into a frame. What I rode was not so much a true ex-"International" model as a hybrid. The frame had completed the week and so had the 235 c.c. engine—but not in strict company.

The engine between my legs when I found that fifth gear was of overhead-camshaft type. And thereby hangs an interesting tale. The original Lodola—a one-seven-five—was also of o.h.c. design, with the camshaft driven by chain. The job went into production in that form and it was only after it had been in the hands of private owners for some time that snags began to arise. Mechanically clever though the average young Italian is, he ran into trouble after each decoke: he couldn't get the valve timing right.

With their usual efficiency Moto-Guzzi grappled with the snag and devised what they thought was a foolproof solution. But even after that young Tonys found themselves getting the timing wrong and so, reluctantly, the overhead-camshaft was put aside and a low shaft and pushrods were substituted. But when the I.S.D.T. came along Moto-Guzzi saw to it that all eight models entered were of o.h.c. type.

In that form, as you will know if you were down in Wales, the model is quite magnificent. Give it to any of our own Trophy lads and he would say the same. For my outing that fine morning I had with me an Italian Trophy man, Pino Radice, on another scratch model. They don't take chances at Mandello.

Before the gallop on the road I was able to indulge in some mild rough stuff. *Before riding on the road, note, not after.* But there were no problems, none of the strangeness that sometimes comes when you straddle an unfamiliar model.

Probably never before have I felt so immediately at home on any motor bike. I prodded it to life, headed it at a steep climb and did a Jeff Smith from the summit. It was as simple as that. The model landed back wheel down, dead square and with next to no sensation of landing.

Down a gear, smartly round left and a wall-of-death along a 45-degree bank: the Lodola stayed effortlessly up on the camber so long as the power was turned on. When the taps were eased we left the bank and rejoined the flat for another series of loops and jumps, with landings both on the flat and downhill, and more up-and-off-the-bank stuff.

On the left George Wilson chats with the famous Moto-Guzzi technician, Ing. Giulio Carcano, and Sig. Todero of the design staff. Wearing the helmet is Pino Radice. Right: "I was reminded afresh of the T.T. Guzzis." This is a 1949 shot of Wilson on Barrington's Lightweight-winning model

MODELLO SPORT

International Six Days Lodola and a Stornello Sport 125

The handling was fantastic. It was light but not over-light. It was true on ruts that ought to have deflected the front wheel. Both front and rear suspensions were so unobtrusive that one might have been riding on air. That handling further proved itself on the lake road, and when Pino later led me up the local mountain.

Let me tell you about that mountain road. It is very narrow—of course!—little more than the width of an Italian bus (as I was to discover when we met one on the way down). The surface alternates between loose, ball-bearing stuff and sections of smooth bitumen. The innumerable hairpins are cobbled.

But these hairpins are never dry. Water trickling down the mountainside runs across them and carries with it a film of mud. Crossing these rivulets Pino was masterly: crank hard over into the hairpin, lift the wheels as near to the vertical as possible on the goo, and then over again and the taps opened hard.

The five gear ratios were perfectly calculated for going of that type. Bottom was so low that it could be reserved for climbing the sides of houses. Second, third and fourth were so close that, particularly between the upper pair, you could quite easily "lose count" and wonder which gear was, in fact, in mesh. Top was overdrive.

This will give you some idea: gearbox ratios are 3.25, 2.28, 1.81 and 1.44 to 1. Top is of course, direct, providing an overall ratio of 6 to 1, which is high for a hotstuff two fifty. (The top gear of the Honda Super Sports tested last week is 6.68 to 1; both models have 18in-diameter wheels.)

The engine would rev and rev, then rev some more, so much so that valve float was very difficult to pin-point. And the model had neither speedometer nor rev-meter! What the genuine maximum was I therefore don't know, but that it was more nearly 80 m.p.h. than 70 I am certain.

Hotstuff the Lodola certainly is. Up that mountain it could, at any instant, be going as fast as I, at any rate, wanted to be going. The way it streaked away from hairpins was proof enough that though Moto-Guzzis might have retired from racing they haven't put all their textbooks away. I was reminded afresh of the T.T. Guzzis I used to sample in the Isle of Man.

In spite of its ability to climb the rev scale the Lodola doesn't rely on revs for power. In true Moto-Guzzi fashion it has an outside flywheel—and the inertia effect of that additional steel disc undoubtedly contributes vastly to the smoothness of power delivery at low r.p.m. The drive was entirely unobtrusive. As for the riding position and control layout: you just couldn't ask for anything better.

I wish I could say the same about the riding position on the Stornello Sport. It is just about as ridiculous, for me, as anything could be. I could have sat on the Lodola all day without the slightest discomfort. After four kilometres on the Stornello I had a pain in the neck (both literally and figuratively) and after ten I had a pain in the back as well. How simply a potentially magnificent motor cycle has been spoiled, for the little job has ample punch and handles like a thoroughbred.

Basically the model is very similar to that we know from *The Motor Cycle* descriptions. The main differences are that whereas the standard model has parallel overhead valves and a flat head the Sport has a hemispherical head, inclined valves and a high-compression piston (the ratio is 9 to 1). The piston, by the way, has an additional oil control ring operating in a groove in the skirt—a feature which permits larger-than-usual oil circulation without any increase in consumption.

For a sportster the job is remarkably quiet both mechanically and on the exhaust. I couldn't help reflecting, too, how much more quiet it could seem to the rider if he wasn't sitting with his head plumb over the rocker box. As I mentioned earlier, handling also is out of the top drawer, and that could be even better if a more upright posture were possible and less of the rider's weight placed on the handlebar as with the standard model.

Above is an I.S.D.T.-type Lodola in its Milan Show setting. The model has won the Italian Regularity Championship

Right: The Stornello Sport 125 impressed for its quietness, power, handling and braking—but not for its riding position!

110 cc MOTO-GUZZI ZIGOLO

EASY on the eye, fun to ride, yet not a bank-breaker when it comes to sordid finance. That sums up the 110 cc Zigolo. The machine is unmistakably Italian in design. Unlike some of its contemporaries from that country, however, it has a specification which appeals to a wide cross-section of riders ranging from the impecunious youngster to the old hand needing a sound lightweight.

Coming from the Moto-Guzzi factory, the Zigolo inherits a tradition of fine engineering and race-bred handling. The first quality is achieved in this particular instance without making too many concessions to conventional thought.

A feature long associated with the marque is a horizontally disposed cylinder for the two-stroke engine. Mixture enters the crankcase by a rotary valve on the right. The connecting rod is attached to an overhanging crank. Further proof that the engine design is up-to-the-minute is the recommendation that a 50 to 1 petrol mixture is adequate.

A pressed-steel body and rear mudguard unit is attached to a tubular backbone with location points for the steering head and rear-fork pivot. On each side of the power unit, quickly detachable, light-alloy covers permit ready access to the carburettor, contact-breaker points and gear-box filler.

Although the Zigolo is a small capacity, lightweight mount, the riding position provides a comfortable handlebar-seat-footrest relationship. Light-alloy clutch and front brake levers operate on pivot blocks welded to the handlebar. Finger adjusters are fitted for cable adjustment.

Gear-changing is by a rocking pedal which can be operated with the heel or ball of the foot, using the footrest as a pivot. Rear brake operation is by means of a pedal which, although non-adjustable was suitably placed.

On the left of the handlebar a neat unit houses the horn button, and light and dip switches.

A hinged door on the near side of the body covers a tool compartment. On the offside, a similar door gives access to the fuel tap.

Contrary to British practice,

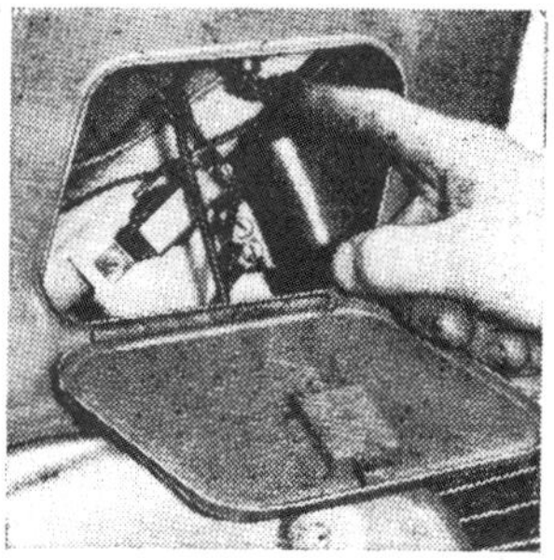

Above: The spring-loaded fuel tap is concealed behind a hinged panel on the right of the body pressing. Below: Speedometer drive is from the front hub

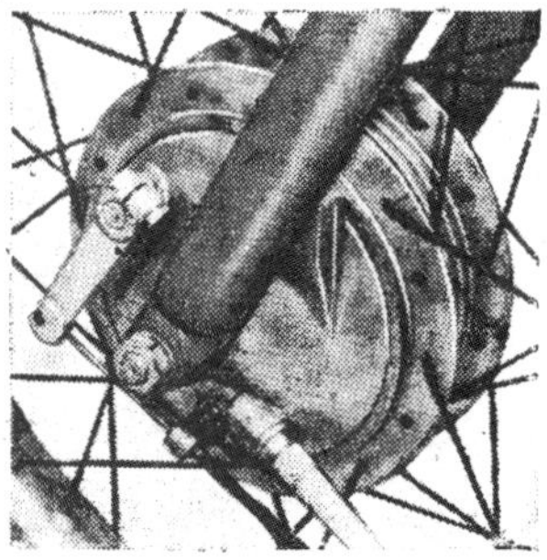

the kick-start pedal is on the left. One quickly becomes familiar with the layout and only a couple of light prods were needed to bring the engine to life from cold. Indeed, with first gear engaged it was possible to paddle away astride the model, drop the clutch and go.

Excellent two-stroking and exemplary smoothness of the power unit at high or low engine revolutions are two outstanding features. The very low mechanical and exhaust noise level makes the Zigolo a most acceptable mount in built-up areas. (One has to listen closely to detect the faint hum from the helical-gear primary drive.)

The wide spread of power makes the fact that the gear box has only three ratios less obtrusive, but a fourth would be preferable. Quiet gear changing could usually, but not invariably, be achieved by slow, deliberate pedal movement. Apart from the click, the gear change was faultless.

Light and smooth, the clutch stood up to all the hard work it was given without any tendency to lose its adjustment. Ridden one-up, starting from rest on a one-in-six gradient was a practical proposition without the need for excessive clutch slipping.

The clutch freed completely —there was never any difficulty in selecting neutral with the engine running while the machine was stationary.

With the horizontal engine layout giving a commendably low centre of gravity and a frame free from whip, good handling would be expected. It is first class in every way, with properly damped front and rear springing ensuring that there was no pitching under any conditions. Furthermore, suspension bottoming was never experienced.

CRUISING

Brakes were adequate and, with the excellent handling, enabled full use to be made of the performance. A happy main-road cruising speed for the engine was between 40 and 45 mph. (The speedometer read approximately ten per cent fast through the range.)

It might appear that its low, forward position would make the sparking plug extremely vulnerable in wet weather despite the waterproof cover. To test this the Zigolo was ridden at speed on unmade farm roads abounding in puddles, some of them about four inches deep. No trouble whatsoever was experienced.

In fact, the front mudguard (and the rear) was most effective. Very little water was thrown up on to the main bodywork.

Direct lighting is fitted. The headlamp beam was powerful enough for speeds of up to about 40 mph at night. The dip switch is a little awkward to use with a gloved thumb and was inclined to be sluggish in operation. A louder horn would be an improvement.

With its sound engineering and excellent all-round performance, the Zigolo is a really gone guy.

SPECIFICATION

ENGINE: Moto-Guzzi 110 cc (52 x 52 mm) single-cylinder two-stroke with rotary inlet valve. Light-alloy cylinder head; light-alloy cylinder barrel with chromium-plated bore. Overhung crankshaft supported by one ball and one roller bearing; needle-roller big-end bearing. Compression ratio, 7.5 to 1. Petroil lubrication; ratio 50 to 1.
CARBURETTOR: Dellorto, with air filter. Air slide operated by handlebar lever.
IGNITION and LIGHTING: Filso flywheel generator with lighting coils and remote-mounted ignition coil. CEV 5in-diameter headlamp with pre-focus light unit and 25/25-watt main bulb.
TRANSMISSION: Moto-Guzzi three-speed gear box in unit with engine; rocking pedal, positive-stop control. Gear ratios: bottom, 19.4 to 1; second, 11.4 to 1; top, 7.38 to 1. Multi-plate wet clutch. Helical-gear primary drive. Final drive by ½ x 0.305in chain with guard over top run. Engine rpm at 30 mph in top gear, 3,300.
FUEL CAPACITY: 2½ gallons.
TYRES: Michelin: 2.50 x 17in ribbed front, 2.75 x 17in studded rear.
BRAKES: 5½in diameter front and rear; finger adjusters. Light-alloy, full-width hubs.
SUSPENSION: Telescopic front fork, hydraulically damped. Pivoted rear fork controlled by hydraulically damped spring units.
WHEELBASE: 49½in unladen; ground clearance, 6in unladen.
SEAT: Saturnia dual-seat; unladen height, 30½in.
WEIGHT: 175 lb, fully equipped and with half a gallon of fuel.
PRICE: £95 19s 6d including British purchase tax.
ROAD TAX: £1 a year.
CONCESSIONAIRES: Pride and Clarke, Ltd, 158, Stockwell Road, London, SW9.

PERFORMANCE DATA

MEAN MAXIMUM SPEED: Bottom: 19 mph
Second: 31 mph
Top: 52 mph
HIGHEST ONE-WAY SPEED: 55 mph (conditions: light side wind; 12-stone rider wearing two-piece suit and overboots).
MEAN ACCELERATION:

	10–20 mph	20–30 mph	30–40 mph
Bottom	3.4sec	—	
Second	4.2sec	5.4sec	—
Top	—	8.2sec	8.6sec

Mean speed at end of quarter-mile from rest: 49 mph. Mean time to cover standing quarter-mile: 26.2sec.
PETROIL CONSUMPTION: At 30 mph, 105 mpg; at 40 mph, 86 mpg.
BRAKING: From 30 mph to rest, 32 feet (surface, dry tarmac).
TURNING CIRCLE: 12ft.
MINIMUM NON-SNATCH SPEED: 13 mph in top gear.
WEIGHT PER CC: 1.59 lb.

Smooth lines make cleaning the Zigolo an easy chore

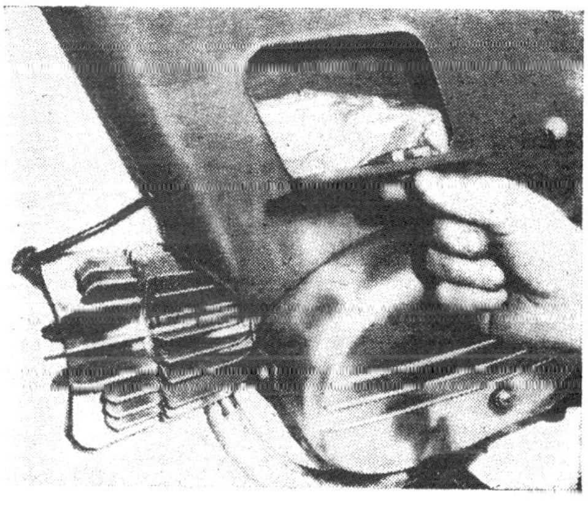

Above: another hinged panel covers a tool compartment of useful capacity

The Most Fabulous Racing Motorcycle Ever Made—

THE MOTO GUZZI V-8

BY CARLO PERELLI

Quietly seated in the meadow near the famous "parabolic" bend of the Monza autodrome, Ing. Giulio Cesare Carcano, the valiant chief of the Moto Guzzi experimental and racing department since pre-war days, was watching the 500cc class riders practicing for the 1954 Italian Grand Prix on a fine September day.

That year hadn't been a very good one for Moto Guzzi in the 500cc class: the four-cylinder in line with indirect injection and shaft drive, designed two years before by Ing. Giannini of Rome (also responsible for the pre-war "Rondine," later to become the Gilera 500cc four) was proving unreliable and difficult to ride, especially on twisty going, because of the notorious torque reaction common to shaft drive bikes. On the other hand, the second racing model fielded by the Mandello del Lario factory, the horizontal "single" developed through the years by Ing. Carcano and his team of brilliant assistants, although excellent on twisty courses, lacked the top speed necessary to challenge the Gilera and M.V. four-cylinders over the fast circuits. But Ing. Carcano still had much faith in the "single lunger" and had already made many plans to improve it for the next season.

So, while lying in that meadow, Ing. Carcano's attention was attracted by a rider on a four-cylinder machine who was making a big riding mistake while negotiating the bend and, even worse, didn't show any sign of improvement with the passing laps. That mistake was repeated at each bend. Actioning his stop-watch, Ing. Carcano registered for curiosity the lap times of that rider: these were far better than the ones set up by the Moto Guzzi riders who negotiated that bend (and all the others) with perfect technique, flat out to the limit! Surprised by this ascertainment, Ing. Carcano said to his assistant: "That's too much. The four is beastly to ride and it's now useless to make any more efforts trying to improve the single. We must start with a clean sheet of paper and design a new multi-cylinder job."

And so it was. The four-cylinder model was scrapped at the end of that season and work started on a highly unorthodox-engined mount, which was later to be known as the "V-8". The new machine featured eight cylinders arranged in two blocks of four, angled at 90 degrees. The axis of the engine was transverse to the frame, to easily fit a chain final drive as in the ordinary motorcycles (more practical and causing no power transmission troubles). A double ohc timing gear system was chosen.

The reasons for this complex engine

V-8 with and without shell. Note central frame tube which doubled as oil reservoir.

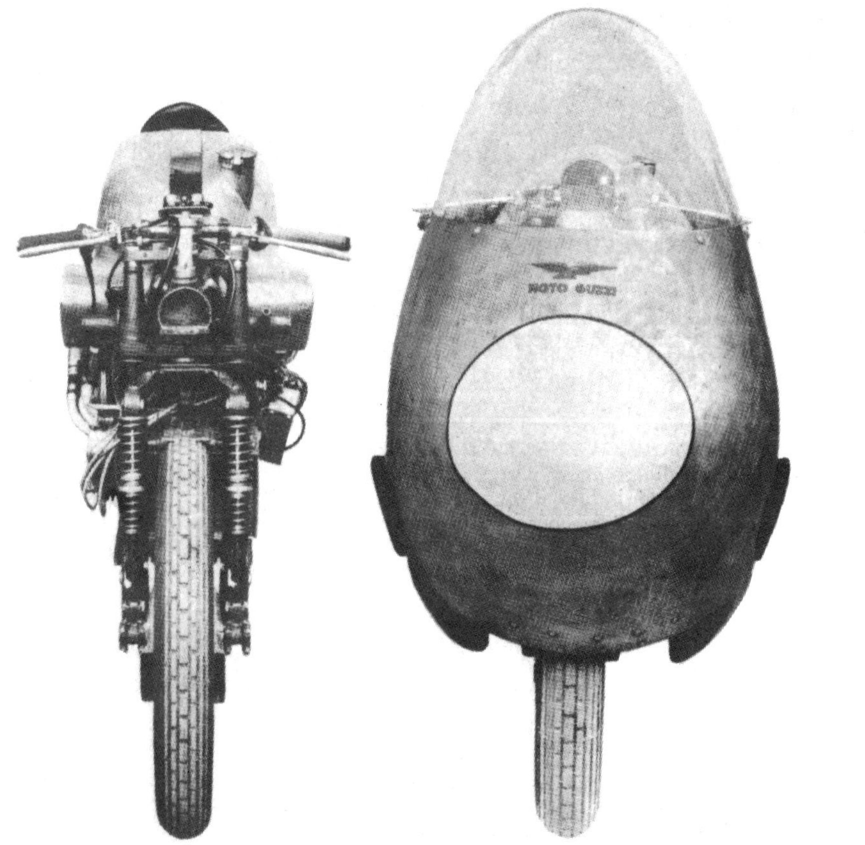

Dickie Dale winning the 1957 Imola Gold Cup on the Moto Guzzi V-8.

design? Cubic capacity fractioned to the maximum for very high rpm and consequent high power output (this was the main target of course), originality of design with great possibility of development in a new field (nobody had ever attempted such an arrangement) and power unit overall dimensions – both longitudinally and transversally – still quite reasonable, enabling the use of a "normal" double cradle frame.

The project was readied in the winter of 1954/1955 with great secrecy and then Ing. Carcano, who has a brilliant, versatile (he is a yacht champion and has improved his boats with ingenious features) and sometimes funny character, sent to friends and the press a drawing of his new machine seen from the right hand side, asking them to guess what it was. Nobody guessed the truth and many strange things were written by the specialized press about it; somebody even ventured to state it was a turbine powered machine! First outing of the "V-8" was during practice for the 1955 Senigallia circuit in Italy. Of course the machine caused a great sensation, but it didn't race; and the same happened after a second appearance. Main teething troubles were encountered with carburetion, ignition and crankshaft lubrication.

Bore and stroke dimensions were 44 x 41 = 62.31cc x 8 = 498.48cc and the engine was thus an "oversquare" with a .932 bore to stroke ratio. In the early stages of development, power was about 65 bhp at 12,000 rpm but gradually this figure grew and at the end of 1957 season when Moto Guzzi unfortunately decided to withdraw from G.P. racing (together with Mondial and Gilera), it was very near the 80 bhp mark at 14,000 rpm; *i.e.* about 10 bhp more than the most potent 500cc four cylinder, and still unsurpassed. It is remarkable that at 14,000 rpm the linear speed of the tiny pistons was still lower than that of a single cylinder 500cc engine (like the Norton Manx or the G50 Matchless) turning over at 7000 rpm!

Owing to its arrangement, the engine required water cooling and the radiator was placed in front of the crankcase. The water pump was driven by one of the timing gears and there were six of them (set in an oil bath case on the right hand side of the engine) to operate the twin ohc for each band of cylinders. At the left hand side of each inlet camshaft were the distributors with four contact breakers each for the ignition. The coils, one for each cylinder, were mounted in banks of four each at the sides of the frame front down tubes. Twin 6V batteries were carried on either side of the rear wheel and the firing order was 1, 8, 3, 6, 4, 5, 2, 7. To have more space for the valves in the hemispherical combustion chambers 10mm spark plugs were mounted.

The very short and massive, forged, nickel-chrome crankshaft with circular flywheels was supported by five caged roller bearings, the races between each throw of the crankshaft being split. Split roller bearings were also used for each of the 90mm long connecting rod big ends. The crankcase was a wonderful one-piece casting in elektron and the cast iron "wet liner" cylinders were finned at the top, within the water jacket, in order to improve cooling. The valves, working in split guides, were set at an included angle of 58 degrees to one another, their diameter being 23mm for the inlet and 21mm for the exhaust, with two coil springs each, over which the large tappets were fitted in direct contact with the cams. The domed pistons, deeply recessed for valve clearance and also to get a squish effect, carried two compression and one scraper ring. Lubrication was by dry-sump system, with double gear pump driven at half engine speed, old traditions for Moto Guzzi racers; another tradition, but more recent, was the oil tank (capacity 9 pints) in the large diameter top frame tube.

Each cylinder was fed by one 20mm Dellorto carburetor and two float chambers on the left hand side served four carburetors each. A single cable from the

Moto Guzzi V-8

twistgrip operated the eight throttle slides with a system of levers which reduced effort to the minimum. The multiplate dry clutch was located outside the crankcase on the left hand side and primary drive was by means of two spur pinions in oil bath giving an engine to gearbox ratio of 2.75:1. There were four, five and even six-speed gearboxes (always in unit with the engine) to be employed according to the various circuits. But it was found that, thanks to the tremendous pulling power of the eight cylinders from low rpm ranges, the six-speed type was practically useless. Incidentally, this high torque feature of the engine at low rpm was very useful on twisty circuits and so the bike was equally at ease on every type of racing ground.

The frame was of the double cradle, tubular type with the pivot point of the rear swinging arm incorporated in the crankcase casting to reduce the machine wheelbase. As with all post-war Moto Guzzi racers, front suspension was by mean of leading links, with exposed shock absorbers. The front brake had four "shoes" (two on each side, like the M.V. 500) and tires were 3.00-19 (front) and 3.25-20 (rear). In the later stages of development it was found that an Avon rear type tire fitted to the front wheel improved the road holding further. The "anatomically" shaped light alloy fuel tank had a capacity of 8 gallons and the weight of the machine, with full light alloy streamlining, was no more than 319 lbs. Top speed was believed to be around 168 mph!

Its racing debut was made in the 1956 Imola "Gold Cup," ridden by the Australian Ken Kavanagh. The exceptional qualities of the machine were at once evident. In spite of wet weather, the Australian was in the lead for the first six or seven laps recording the record lap in this stage, but he was then stopped by a broken water temperature thermometer. Anyway, the 1956 season was still considered a development one and successes were few, but next year the "V-8" started some successful record breaking. Without special tuning or the use of dope fuels, Bill Lomas dashed along the Terracina (Rome) straight and set up a respectable 244 kmph (151.4 mph) figure for the 10 km (6.21 mile) record, still unbroken. Italian rider Giuseppe Colnago won the 500 Italian Championship and Dickie Dale took the Imola "Gold Cup" race, beating all the best men and machinery of those days. A dolphin fairing was prepared for the twisty circuits and Dickie Dale used it for the Senior T.T. where, in spite of ignition troubles, he finished fourth. But on the ultra-fast Francorchamps circuit for the Belgian G.P. the full type streamlining was employed again and Australian Keith Campbell proved the tremendous possibilities of the V-8 : after the fastest lap of the day at 190.130 kmph (118 mph) he thrashed all the opposition only to retire with a minor snag when comfortably in the lead (a broken wire from the battery).

Racing accidents to the best men on its team prevented Moto Guzzi from fielding the V-8 at Monza at the end of the 1957 season. It would. have been a wonderful swan's song because this machine was superior to all the others although the factory bosses, in great secrecy, had already decided to abandon G.P. racing.

So the most fabulous racing machine ever built, at a cost of millions of Lira, wonderful proof of Italian technical skill, was never to be seen and heard (and a fascinating noise it was) again on the world's most famous racing grounds. All the Moto Guzzi racing material from those glorious days was stored away in a factory shed at Abbadia Lariana, on the shores of Lake Como, and the occasional visitor who can manage access to it is melancholy at the sight of deep dust covering the Moto Guzzi V-8, along with a growing rust which successfully attacks everything . . . •

The Moto Guzzi V-8 from the timing side (top) and left, or drive side (below).

This drawing designer Carcano sent to friends and press causing wild specula tion, none of which was farther out than the truth.

MOTO GUZZI 125cc "Sport"

Pull off the string and paper wrappings, connect the battery, gas it up, turn the key, kick once—and you are off! This is what the CYCLE test staff experienced with the Moto Guzzi 125cc *Sport,* which had no miles on it at all, after receiving it via air freight from the Premier Motor Corporation of Leonia, New Jersey, the US and Canadian distributor for this Italian make. This "one-kick" theme followed throughout almost the entire road test—depending on whose foot was pushing.

As we announced in our "New Models" section last month, Moto Guzzi has returned to our shores. The first ones arriving are the 125cc models, one of which was our test bike.

Older aficionados of the 45-year-old concern will remember that the product of not too long ago had a cruder appearance what with its horizontal single cylinder surrounded with deep, horizontal cooling fins, many parts cluttering up the engine area, and, of course, the exposed flywheel. It was not beauty that sold them but they were very reliable and a real "work horse." A glance at these pictures will tell the reader that things have changed enormously. The cylinder is now slightly inclined, the flywheel is now inside, separate seats for driver and passenger have been blended into one and the clutter has been removed. Now it is one of the cleanest-looking cycles anywhere, and the factory nestled near Lake Como, in the mountains of northern Italy did not have to enclose the bike like a scooter to achieve the effect. Between carburetor and crankcase is a lot of room. A report of the Moto Guzzi models and engines in the fifth issue of CYCLE ever published, back in 1950, had this to say about the engine:

"The Moto Guzzi firm pioneered the horizontal, single-cylinder design of engine. Theory suggested that single-cylinder power impulses could be more effectively dampened if the direction of the impulses were in line with the wheelbase, not at right angles to it. Theory was tested and approved sound. Moto Guzzi have featured this example of imaginative engineering throughout their full line."

Perhaps something happened to the theory or sales appeal of the exposed flywheel design, et al, for these 125's for 1966 are different.

The overhead-valve engine of the *Sport,* with a 9.8 to 1 compression ratio, puts out 12 horsepower. Both the cylinder and head are made of light alloy. The cylinder—tilting 25 degrees from vertical—contains a special, cast-iron inserted liner.

There is no support under the powerplant so it really becomes part of the frame. Four nuts secure the crankcase to two, slender down tubes in front, and another four nuts hold the rear of the engine to the bike.

In this type of engine, the oil combines the functions of a lubricant and a cooling medium. A gear pump draws the oil from the crankcase sump, and it is fed in the crankshaft after being cleaned by the centrifugal filter fitted on the crankshaft itself. It then emerges from suitable passageways under the big end which it lubricates, and is flung by centrifugal force on the wrist pin, piston, cylinder walls and gears in the gear box which it cools and lubricates. Surplus oil falls down in the bottom of the crankcase where it collects in the sump.

The rocker-valves assembly is lubricated by oil fed by the pump through a passage drilled in one of the cylinder-head hold-down studs. The oil flowing through this hollow stud falls on the rocker pins and, through suitable drillings, emerges on the rocker pins and bushings which it lubricates. Excess oil falls down in the crankcase well, which is finned outside for better cooling.

The oil level is checked by means of a dipstick, welded on the filler cap, on which are marked maximum-minimum levels. The accessibility of the filler and drain holes make it easy to add, drain, and check the oil.

The air does travel quite a distance before entering the engine. The air is sucked through a short, metal, rectangular tube which leads into the large battery compartment, located just under the front of the seat, then through a ram tube into the carburetor. This design prevents air intake noise, and offers some filtering.

A very long muffler cuts down the air noise in that direction pretty well, while giving the bike that "Italian" sound.

Checking the battery water level is a major operation, and prob-

MOTO GUZZI 125cc "Sport"

ably will be purposely neglected by the average rider of the Moto Guzzi 125 *Sport* for one reason: the heavy rubber band that holds the battery in place. If the rubber was longer or weaker, the two hooks attached to its ends could be put in place without effort. As it is, brute strength and ingenuity are necessary to do the job. We succeeded by pushing a big screwdriver down on one of the hooks with all our might till it caught in place. That band should be changed. In addition, it would be helpful if the battery was encased in clear plastic since it is now necessary to take the battery out of its nest to fill or check the water.

Lying next to the battery in its compartment is the ample and compact tool kit. The allen wrench, oil drain cap, screwdriver, five different spanners, and two tire levers are kept in a heavy canvas pocket.

The owner's handbook is also stored in the battery compartment. Clearly and succintly, the book instructs the owner with the proper maintenance and care of his new possession (and it's in English). One helpful little addition to the photos, to aid the reader in knowing where a part attaches, is the sketching in of hands holding the part near its place on the bike.

Front suspension is telescopic with hydraulic dampers. The rear is swinging fork with exposed coil springs around hydraulic dampers.

Note in the pictures that the lower end of the swinging arm attaches to the frame about five inches in front of the rear axle. If the attachment was made farther back, as is the case with most machines, the angle of the shock would be too severe.

For an easier and quicker change, the rear sprocket is a bolt-on type.

Adjusting both of the brakes to compensate for lining wear is very easy. Only tools needed are a thumb and forefinger. Knurled knobs near the hand brake lever and at the end of the rear brake rod merely have to be rotated to change the adjustment.

Rubber-covered footpegs for the operator fold back and up at a 45 degree angle in case they hit an object, of course. But the other pair, being a few inches higher and back farther, are not as likely to scrape the ground or some immovable object along the way; so they just fold straight up when not in use.

The hands will enjoy operating this Moto Guzzi because they will encircle soft, air-filled, plastic grips that are raised in the center, and pull on clutch and front brake levers with ball ends. With the latter feature, hands are less prone to slip past the knobs; they are a sizable object to hold onto if only a couple of fingers manage to catch the lever in a sudden grasp for it; and the ball ends will not dig into the dirt or hands like the traditional pointed lever can do.

The left thumb has the duty of pushing the horn button and flipping the high-low headlight beam. For either movement, the hand does not necessarily have to leave the grip.

The models we announced in 1950 had the roll-up type stand fitted as do the models 16 years later. But it needs a sidestand for U.S. sales.

The headlight nacelle contains a green neutral indicator, speedometer and the electrical system switch "key." Push the key handle in to turn on the ignition; switch to the right for parking light and left for headlight.

Moto Guzzi should be commended for outfitting their machines with two fine Brevettato products: the gas cap and license plate holder. The fuel tank cap springs open on a hinge when the locking bar is slid over. There is enough room to attach two state license plates to the broad holder, for it measures seven inches wide by six and a half inches high, not counting the tapered section that holds the tailight. The flat, little bars that many other bikes have is conducive to ripping off the plate in the wind.

One thing that mars a clean, simple appearance in a cycle is the mass of wires, cables, and tubes under the gas tank. Not with our test bike. The coil is tucked up underneath the front of the tank and thus right next to the spark plug. Other than this, there are two clear, plastic tubes which transport the gasoline from the main and reserve tanks to the carburetor.

The advertising industry has long known the attention-getting color of red. This is the bright color which stands out on the big, four-gallon tank, in contrast to the black frame and seat, and the gray battery compartment and fenders. A little incongruous, but pretty anyway. Black trim and the company's symbol, an eagle with its wings spreading over the words "Moto Guzzi," also adorn the tank.

All in all, it is a very good-looking bike: in design, detail and finish.

PERFORMANCE

The *Sport* has good low end torque; you can feel its power from dead stop on up. But being a four-cycler, the rider cannot wind it way up in each gear to get more speed and power as is possible with modern, high-performance, two-cycle engines. But, after all, there are four gears at one's disposal and seventy-plus mph is claimed for the top end, so there is really no problem here.

ENGINE MOUNTS in twin down tubes in front. Rectangular pipe under seat is entrance to combination air cleaner/battery box; from there the carburetor draws air through small velocity stack. Coil is mounted directly over the cylinder head. Kickstarter is slippery without a rubber cover.

SWINGING LATCH makes hinged gas cap leakproof and loss proof. Comfortable hand grips are made of air-filled plastic. Appearance of attractive red and black gas tank belies its capacity — four gallons! Green indicator lamp in front of key hole signifies the parking lights are on.

One thing was somewhat bothersome though. The bike seemed to vibrate a little too much, especially for a relatively low rpm engine and rubber-mounted handlebars would help.

Other than this, the ride was comfortable. There is ample cushioning provided on the long double seat for both driver and passenger, and the springing and shocks do their duties admirably.

An important plus for the little machine is the shifting mechanism. The four-speed, foot-change gear box is built in unit with the engine, and is operated with the right foot. We found that we could shift very quickly and smoothly, up and down. The motorcyclist who changes gears while resting his foot on both the heel and toe levers simultaneously will shy away from the Moto Guzzi's arrangement at first. The heel-toe levers are so far apart, it would take a size 12 shoe to keep on with the old habit. Yet there appears to be good thinking behind the "MG" system. When the rider goes through the gears, he is shifting in only one direction at a time; therefore, it is not only unnecessary to be able to quickly push down on either the heel or toe at any instant, but undesirable because he may inadvertently push the wrong one. With the foot setting on only one lever, just one thing can happen: the intended. Furthermore, the feel is more "natural" and secure because the foot can set on the foot rest while either the toe or heel manipulates its own lever—located the proper distance away.

Though firing up the engine was a pleasure, the whole operation could be improved a little if rubber covered the kickstarter pedal. This would cushion the blow if it whanged back against a tender ankle, and would provide a surer footing on the pedal.

Both brakes are full width and internal expanding; the drums are finned. Nothing unusual about all this, but there *is* about the rear brake: It is terrific (as was the case with the rear clamper on another Italian lightweight we tested recently)! On a dime, every time. The front, needing some adjustment, was just so-so.

SUMMARY

The dozen horsepower of the Moto Guzzi 125cc *Sport* does not make it a bomb. However, apparently this brand of motorcycle is *reliable*—a more proper criterion than power for a lightweight street bike. In announcing the 1966 models we said, "Owners of the familiar, old exposed-flywheel models have claimed upwards of 100,000 miles on the odometer without ever having removed a case for a peek inside."

It shifts smoothly and rapidly, stops quickly, starts easily, is good looking and is reliable. That's plenty. ◀

SPECIFICATIONS
MOTO GUZZI SPORT 125cc

Engine	4 stroke, single cylinder, OHV
Cylinder	Light alloy with cast iron liner, inclined 25°
Displacement	125cc
Bore	52mm
Stroke	58mm
Compression ratio	9.8:1
Horsepower	12
Engine lubrication	Oil sump in crankcase
Oil capacity	½ gal.
Oil filter	Screen type in crankcase, centrifugal on crankshaft
Carburetor	Dellorto ME 18 BS
Ignition	6 volt, 28 watt flywheel alternator with external H.T. coil
Battery	6 volt, 7 amp
Electrical equipment	Headlamp contains key which operates ignition and lights; dipper switch and horn button on left handlebar
Clutch	Wet multi-disc
Transmission	4-speed constant-mesh
Gear ratios, overall	25:1, 14.9:1, 10.6:1, 8.2:1
Drives	Primary by helical gears; secondary by roller chain
Front suspension	Telescopic fork with hydraulic dampers
Rear suspension	Swinging fork with coil springs in hydraulic dampers
Tire sizes	2.50x17 ribbed (front), 2.75x17 R block tread (rear)
Brakes	Expanding; right hand operated (front), left foot operated (rear)
Wheelbase	50"
Overall length	76"
Overall width	22"
Overall height	36"
Road clearance	6⅜"
Weight	203 lbs.
Fuel and capacity	High octane, 4 gal.
Fuel consumption (claimed)	89 mpg
Climbing ability (claimed)	31°
Maximum speeds (claimed)	20.5 mph (low), 34.1 mph (second), 47.8 mph (third), 70 mph (fourth)
Price	$429, East and West Coasts

CONVENTIONAL SWING-ARM with forward leaning shocks make up rear suspension. Knurled alloy knob makes for easy adjustment. Pirelli tires mounted on steel rims do the ground work. Close inspection reveals the fine paint job on frame and fenders. Exposed springs make for a racy appearance.

ART DIRECTORS Paul Halesworth (left) and Jim McGowan enjoying the stable and comfortable ride of the Moto Guzzi "Sport." But the most noteworthy thing about the machine is its reliability.

CYCLE ROAD IMPRESSIONS

MOTO GUZZI 125 I.S.D. TRIAL

That's ISD not LSD, and this sturdy, high-strung motorpsychle tripped through our "acid" test muddy but unbowed.

A thoroughbred is the result of selective breeding – high-spirited but temperamental. So it is with the Moto Guzzi 125cc ISD Trial Motorcycle—an infuriating/endearing, high-strung love object; but a true trail-breaking offspring from a long line of fine off-the-road performers.

To start... was the problem. It seemed that our little motorcycle was determined to be an all-range, short or long throw, non-environmental spark plug fouler. It would hardly start with a run, bump and heave. But three plugs, a full tune-up and much gutter language later, a feeble cough shook the frame and it burst into its throaty song. In fact, after a few weeks the problem was almost forgotten; and while never an easy starter (11.4 to 1 compression ratio, you know) the initial effort no longer overshadowed all the bike's good qualities.

The starter pedal also held our enthusiasm to controllable levels. Apparently designed by a sadistic misanthrope, it was well within bruising range of the foot peg, leaving an unhappy reminder on our collective editorial shin.

A must when sparking off this model is judicious use of the carburetor primer and two gas petcocks. If you want to start again, turn off the petcocks the minute you stop, or you might begin to feel like Noah.

The ignition rod (one hesitates calling it a key) is another oddity. You drop this little Bakelite and metal device into the hole in the headlight (any old position will do, although a turn to the left actuates the lamp) and contact is made. Not bad, except that moderate bumps tend to jar the rod loose, contact is lost, and we play the start game again.

But when you get it going, the Guzzi will take you anywhere you'd ever want to go. We spent a weekend in the mud, a weekend in the woods, and a weekend on snowy sand and gravel, putting the bike through its paces. (Let the fact that we thought enough of the bike to keep it for staff use for over a month speak for itself.) The Guzzi accepted every environment we found for it happily.

A lot of the bike's success can be attributed to the weight (a mere 208 pounds) which is a definite asset in soft terrain.

Also contributing to tractability were the great (2.50 x 19 in. front and 3.00 x 19 in. rear) Pirelli "very-knobbies." Yet five minutes riding that first weekend almost obliterated these advantages. One run through the rain-soaked clay basin and the one hundred-fifty pound rider weighed one-ninety; the bike two hundred-seventy and the tires looked like drag slicks. Still, the Guzzi careened on . . . out of gullies, over hills, through streams and across great furrows (dug by a well-meaning but out-of-place "Street/Scrambler").

The Guzzi is no plodder like many trials bikes. Even in the rough, it's flexible, strong and smooth, prancing through the first three gears. The very strong gearbox and clutch, by the way, shifted smartly and always took up cleanly, despite the abuse. A ride on the road gave the same impression until we hit fourth gear. With its 10.25 to one ratio, fourth gave the impression that a tail wind should be optional equipment; it is much too tall.

If you are planning on doing a lot of highway driving, there is a Moto Guzzi work horse of the same displacement which is made for the street. Called the 125cc Sport, it's an entirely different breed whose reliable 9.8 to one compression ratio yields a plugging 12 horsepower and a reported 70 mile-per-hour top end. Use this street model when you ride the road; leave the plucky model for the terrain for which it is exquisitely designed.

Interestingly enough, our sensitive Trial model has 13.5 hp at 6,500 rpm. Quite a reversal in the motorcycle industry, where fine tuning usually goes to the street version. Maybe a vestige of the European ISDT high-speed sections.

The telescopic front fork with hydraulic oil dampening gives almost six inches of travel. Even with an inch of mud under the fender, the tire never scraped. The top mounting lugs of the suspension are a bit more forward than we prefer, but the swinging fork with conventional coil springs and dampers do their job well. Rear end wobble never showed itself on our machine, regardless of terrain. The frame is a sturdy welded tubular structure. Front down tubes bolt to

Continued on page 97

PHOTOGRAPHY: LILLIAN LANGOTSKY

NEW MOTO GUZZI 700cc V-Twin

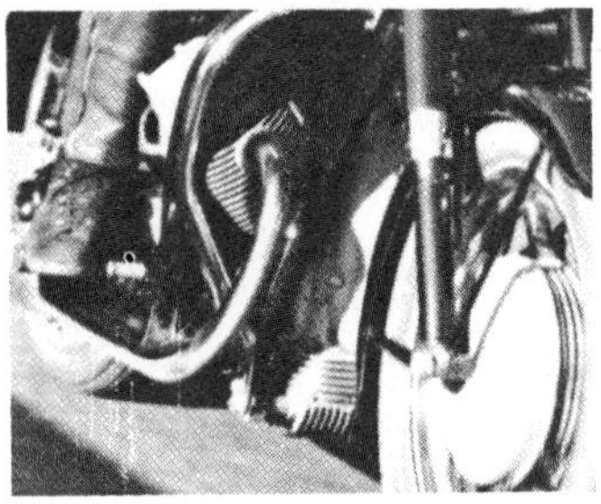

THE Italian motorcycle industry, although fighting against a recession, is showing a revival of interest toward the "big 'uns." Ducati is preparing the huge Apollo 1260cc four-cylinder primarily for the U.S. market, Gilera is considering production of a commercial road-going version of its famous 500cc racing "four," and, as previously announced, Moto Guzzi has built a rather unorthodox 700cc. Called the V-7, it is a transverse V-twin with final drive by shaft enclosed in the right arm of the rear suspension.

After some two years of intensive tryouts, the new Moto Guzzi is now at the end of the testing stage. Mainly intended for army and police use, it isn't much of a sporting mount, but it is a grand touring job, extremely sturdy, needing infrequent maintenance yet offering quite satisfying performance even to the speed enthusiast.

The pushrod-operated OHV engine develops 42 bhp @ 6000 rpm (with terrific pulling power from the lowest rpm ranges), so a top speed of slightly over 105 mph is assured without any sign of fatigue, according to testers.

Ignition is by 12V battery and coils, with generator and contact breakers at the base of the cylinder "V" and belt-pulley operated from the front part of the crankshaft. The electric starter motor, engaging a car-type flywheel, is on the left side of the crankcase. Other interesting features are: single-disc dry clutch, four-speed gearbox in unit, heel-and-toe pedal controlled from the right side, and two large bore Dellorto carburetors. The V-7 has a double cradle tubular frame with orthodox suspension and 18-inch wheels, double leading shoe front brake, single leading shoe rear brake (both light alloy, full width), and weighs 495 lbs. in running order. Riding position is quite high and slightly to the rear, but fairly comfortable.

Deliveries of the V-7 to the Italian Army and Police are expected soon (both have already tested it with excellent results), and afterwards it will be available to the general public. Pictured here is the "civilian" version; we are looking forward anxiously to road testing one! •

Guzzi V-twin features electric starting.

CYCLE ROAD TEST

MOTO GUZZI V-7

An exotic transverse-V engine and the latest in light-alloy technology lend zest and stamina to this newest of the big-bore tourers.

If we had to choose a nickname for the Moto Guzzi V-7, we'd have to call it "The Persuader." Ordinarily this underworld term applies to small revolvers of large calibre, to discreet lead-weighted implements, or to certain physically overdeveloped individuals with quiet voices and one-track minds. Not that you can argue the size of the V-7: first confrontation is certain to be a staggering experience. Our nickname refers more to the way this Italian wonder works on you.

Only after we had ridden the V-7 some 2000 miles and returned it to the brothers Berliner of Berliner Motor Corp. did it occur to us to road test the machine. And then we had to go ask for it again because no one could remember enough about it to write a detailed test report. Call it poor planning—what can we say? We climbed aboard, turned the key, and it started. Roll on the throttle, ease out the clutch, and head for points south—Daytona in particular (see page *78*). A long ride, a fast ride, a trouble-free ride. We were lulled like so many owners of American cars. "It runs, so why look under the hood?"

That, eventually, did the persuading. It ran. It *really* ran. And the carburetors were wildly over-jetted and we were too much in a hurry and too lazy to do anything about it. And still it ran, winging up to 90 and faster in an instant if you happened to forget that they patrol those southern Interstate routes just as heavily as those long stretches of single-lane 301. And maybe we expected it would lumber along like the milk wagon that it, with all its aluminum and chrome, faintly resembles. But it didn't, it sprinted . . . sort of like those 230-pound pro-football tackles who can do a hundred yards in ten seconds flat. From anywhere in the range, in virtually any gear, just twist the throttle and the V-7 takes off, majestically, but in a hurry.

In fact the weight thing ought to be cleared up. The V-7 *looks* as if it weighs at least as much as America's traditional two-wheeled highway heavies. But it ran up our scales, soaking wet, to 540 lbs. Large-displacement touring bikes vary, in general, from 380 to 725 lbs. stock, putting the V-7 in midrange. It looks so heavy because its very bulky engine along with associated running gear completely occupy all the space between the front down-tubes and the rear wheel hub. There isn't even any room for *light* to get through. But what gives the lie to the look is the extensive use of aluminum and light alloy in the engine castings and other fittings. For what it carries, the job

PHOTOGRAPHY: MICHAEL SOLDEN

it's meant for, and the power it generates, the Guzzi is well proportioned in the weight department.

The powerplant is far-and-away the most interesting and most distinctive feature of the Guzzi. It is a 700-cc V-twin mounted transversely in the frame and driving the rear wheel through a shaft and bevel-gear arrangement. This configuration has been tried several times in the history of motorcycling (it was, in fact, the original layout of the famous Brough Superior), but it has always been superseded. Actually, though, most of today's popular layouts were tried and abandoned several times before they finally became "conventional." The primary fault of the early transverse V-twins was excessive vibration—transmitted to the frame at a point (the forward end of the cradle) where it was most prone to resonate with the structure.

Somehow Guzzi's engineers have been able to overcome this difficulty. They have compensated for the unequal spacing of the cylinder firing pulses, found a suitable balance factor for pistons operating at a 90° angle to each other, and countered a slight "rocking moment" from the offset cylinders. Yes, the machine still vibrates, but the vibrations are quite muted and of very low frequency. We were unable to detect any resonance point in the rpm scale—that is, no place in the entire speed range where vibration increased sharply to make the bike feel as if it would shake apart. Such resonances are common in big twins, and all too often they occur at just the speed that one prefers for cruising. In the Guzzi there is just a quiet, low throbbing, as if the rider is sitting on the bridge of a steamer with the engine room a long way below-deck. No spine jangling, no buzzing, no fillings rattled loose, no hands or feet going to sleep or turning numb. In a word, smooth.

The engine is tall and that might lead you to think the bike's center of gravity is high, adversely affecting handling. Again not so, for the tall housing is mostly lightweight alloy. The heavy pieces, like the steel crankshaft and the flywheel,are down low. And the engine unit sits deep in the double cradle frame with a delightfully large (three quart) oil sump at the bottom.

One key to understanding this engine's virtues is knowing that it formerly powered a small auto used by NATO forces in Europe. It was then fit into a motorcycle for military use in Italy. Two degrees of tune are assembled by Moto Guzzi, one with 7.5:1 compression and a peak bhp of about 38 at 6000 rpm. The other, with 9:1 compression and more sporty valve timing, rides in this V-7. Both military applications demanded what the touring motorcyclist wants: reliability. A 704 cc engine generating 60 SAE horsepower at 6000 rpm actually delivers about one bhp per cubic inch or 60 bhp per liter at the rear wheel. These numbers have long been the rule-of-thumb limits for long-term reliability. A bore and stroke of 80mm by 70mm indicates very low mean piston speed. Even the cylinders, which are cast aluminum alloy with hard-chrome linings, are clearly destined for durability.

Features common to automobiles abound on the V-7. The crankshaft turns in automotive type, aluminum-tin plain bearings and the steel connecting rods have thin-wall racing shells of the same material at the lower ends. Stressed as they are in this relatively low rpm engine, these bearings should prove quite adequate over a long period of time.

Placing the engine and drive components all in a line in automotive fashion usually takes a heavy toll in serviceability, as anyone who has ever taken his car apart knows. But the engineers at Moto Guzzi have also miraculously sidestepped this snag. Unlike most shaft-driven twins, the V-7 is quite easy to get apart, service, and put together again in a short time.

The top end, of course, hangs right out there in front of you, making valve and cylinder jobs a snap. The carburetors are large and well tucked in, but the mainjets and adjusting screws have been made quite accessible. Even removing the entire engine assembly is not all that difficult. Detach wires and cables, take off the tank, undo two through-the-frame bolts, the drive-shaft joint, and then just roll out the engine, complete with transmission and starter.

Not that all of this is likely to be necessary, at least before 60,000 miles or more. And then, you'd only have to do it if the mains were gone. Mainbearings are factory pressed into replacement-unit housings, which bolt to the inside of the crankcase shell.

Replacing the (slightly more vulnerable) lower-end bearings can be done right through the bottom of the crankcase without removing the engine from the frame at all. Just drain the oil, unbolt the pan (just as you would with your car), and you can easily reach the connecting-rod bolts. These undone, you need only peel out the shells, replace them with new units, and bolt it all back up again. An entire engine job takes only one special tool, an oil-pump puller.

The most conspicuous automotive feature on the V-7 is its Marelli starter motor geared to the outer rim of the flywheel-clutch housing. There is also a belt-driven, 300-watt dc generator, a remote (driven through a bevel gear on the camshaft) automotive ignition distributor with automatic advance, a powerful, gear-type oil pump, and a large-diameter dual-plate dry clutch. A final touch of automotive irony is that no provision for kick starting has been made. Moto Guzzi

figures that if the Marelli starter motor worked successfully in over two million Fiat cars, it will bear up in the V-7. If your battery goes dead, you *can* bump start the V-7, but you won't enjoy it.

Behind the engine sits the Guzzi's bolt-on gearbox. It is slightly off-center to make way for the starter motor, and the shift rocker-pedal is on the right. The pattern is one up and three down with the toe, but we quickly got used to the heel-and-toe rocker. Ratios are well chosen and the box has a cam-and-spring internal shock cushion to preserve those gears. The only fault we could find with this sturdy and somewhat deliberate unit holds also for the bike's rear drive hub: a seeming lack of breathers or vents. Maybe we didn't see them, or maybe they've got sealing perfected to obviate them. In any case, the gearbox, rear-drive unit, and crankcase (which *is* vented) remained oil-tight during the 2000 miles we put them through.

The frame and suspension on the Guzzi are more or less conventional by present day standards. The large, double-loop cradle joins a single backbone tube of enormous cross section (and consequently, strength). The sub-frame for mounting seat and rear-suspension points continues almost to the rear of the machine, providing a firm anchor for racks, saddle bags or other accessories you might want to fit. Already included is a broad and comfortable dualseat (from Italy, no less!) and two sturdy steel hand-grips for your passenger. There is also the usual passenger grab strap. The rear spring-and-shock-absorber units have no covers, but are chromed nicely and are adjustable to three degrees of tension.

Up front sits a *very* heavy looking telescopic-fork unit. Maybe it's just fat looking, along with its 4.00 x 18 front tire. Here again, appearance is deceptive, for the V-7 doesn't steer anything like a milk wagon. On the contrary, the front end, once the bike is underway, handles with remarkable ease and lightness. You don't have to point that big tire anywhere, just lean into the turns as you would on a lightweight. This lightness is an extraordinarily pleasant feature on a motorcycle of such size, and other manufacturers might take note.

Most of this good handling owes to good choice of fork rake and trail, but some arises from the machine geometry induced by a transverse-V engine. With those cylinders out there cooling splendidly in the airstream, the rider must sit slightly further astern for adequate knee clearance. This shifts the overall center of gravity toward the rear. Put a passenger on the dual seat and a mountain of luggage on a rear rack and you're liable to find that featherweight front end wandering a bit at high speeds. In all but this most extreme case you will be delighted by the agility of the V-7. Fork damping is two-way and may be altered by changing the viscosity of the fork fluid. There is not much travel in these forks, but the V-7 is a highway machine. Any owner can easily find the right damping fluid for the type of riding and the roads he will encounter.

One of the more attractive features of the V-7 are its brakes. Here at last is a large, heavy-duty touring bike with brakes adequate for the job. Anyone who has examined our specific brake loading figures over the past few months will notice how miserably most tourers do in terms of load versus brake swept area. The Guzzi V-7's brakes, both front and rear, have full-width hubs and shoes, with a diameter of nine inches. The front unit has twin leading shoes and can be worked hard. The rear is a single-cam version with plenty of stopping power (you can lock up the wheel without he-man pressure), but still useful for holding a loaded bike on a uphill grade while waiting for a light. The drums are aluminum alloy with cast iron liners, all of which minimizes unsprung weight and helps handling.

The Guzzi will stop on a dime and do zero to sixty in a little over seven seconds. It is robust and apparently built to last. Yet it certainly isn't a high-performance play bike for sprinting about the neighborhood. There's a lot of high-response torque in that engine, but not the kind that will lift the front wheel off the pavement at 60. No, this bike will be happiest touring, touring on high-speed turnpikes or down rolling back-country roads. We did four-hundred-mile days without discomfort, and if you can find us a longer-distance fanatic, we're sure the bike will outlast him.

Meanwhile, with its lightweight handling, the V-7 will double nicely for chores and commuting. There's so much to it like the family car: electric start and all that, plus instrumentation that indicates oil pressure, neutral, high beam, and charge. Turn the key one stop, read all the lights, turn it again until she fires over, and away you go. Humble and self-effacing as we are, it took us only about five minutes to learn that the V-7 is a real attention getter. Stop anywhere and people gather around, cyclists as well as the curious. There's a lot of fine detail work there, a lot of gleaming alloy and chrome, a dazzling red tank, a BIG (looking) engine, and an authoritative sound. America has been waiting for a really new big-bike for a long time and here's the Moto Guzzi V-7, the first to step into the ring. Why not climb in and let it try to persuade you? ◉

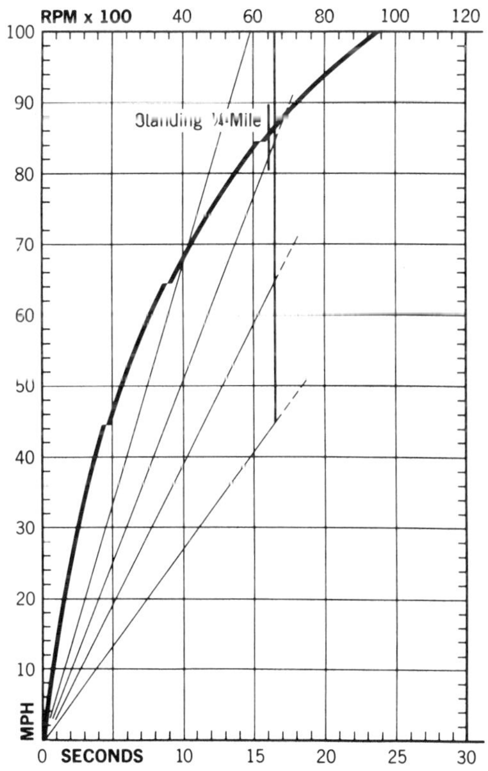

MOTO GUZZI 704cc V-7

Price, suggested retail	East Coast, POE, $1439.00
Tire, front	4.00 in. x 18 in.
rear	4.00 in. x 18 in.
Brakes, front	Twin leading shoe, 9 in. x 2.00 in.
rear	Single leading shoe, 9 in. x 2.00 in.
Brake swept area	113 sq. in.
Specific brake loading	5.95 lb/sq. in.
Engine type	Four-stroke twin
Bore and stroke	3.15 in. x 2.76 in., 80 mm x 70 mm
Piston displacement	42.9 cu. in., 704 cc
Compression ratio	9:1
Carburetion	(2) 29 mm, Dellorto
Air filtration	Paper element
Ignition	Battery and coil
Bhp @ rpm	50 @ 6500
Mph/1000 rpm, top gear	17.0
Fuel capacity	4.5 gal.
Lighting	Generator, 300 watts
Battery	12 v, 32 ah
Gear ratios, overall	(1) 11.33 (2) 7.82 (3) 5.98 (4) 4.57
Wheelbase	56.9 in.
Seat height	31 in.
Ground clearance	5.9 in.
Curb weight	540 lb.
Test weight	715 lb.
Instruments	Speedometer
0-60 mph	7.7 seconds
Standing start ¼ mile	16.1 seconds—84.9 mph
Top Speed	102 mph

A Great Legacy of Exotic Racers

BY GEOFFREY WOOD

MOTO GUZZI HISTORY

IN THE GREAT SAGA of international motorcycle racing there are many colorful chapters, but for pure nostalgia and exotic machinery none can equal the story of the Moto Guzzi. Not a common machine in the Americas, this Italian marque has its home on beautiful Lake Como in northern Italy, and the story of its life is one of the most colorful and intriguing in all the world.

Italy has traditionally been a rather poor country, and the benefits of the industrial revolution were rather late in coming to its people. When mechanization finally came, though, it was carried on with such fervor that Italy's motorcycles became pre-eminent on the European classical road racing courses, as well as being some of the finest in the world for day-to-day transportation. An important part of this rise in industrialization was the legend of the Guzzi — a story that put the Italians on wheels, made the company the best seller in its land, and brought home the prestige of many world championships.

This great legend all began way back in the days of World War I, when Giorgi Parodi and Giovanni Ravelli were flying airplanes in defense of their country and the two flyers' planes were serviced by a young technician named Carlo Guzzi. This particular mechanic could talk about nothing except his dream of building the finest motorbike in all the world. In off-duty hours, Carlo would sketch out his design while the two flyers listened. And Ravelli, being an experienced motorbike racer, voiced the opinion that the design had great merit.

So, after the war, Carlo Guzzi and Giorgi Parodi got together on the project and laid plans for the new G & P motorcycle. The emblem selected for the new make was an Eagle with its wings spread wide in flight. This was out of respect for brother Ravelli, who had lost his life in action.

The first problem was to raise the capital to begin operations, so Parodi went to his father who was a man of some means. The necessary finances were duly obtained (some 2000 lire) and the elder Parodi promised more, if the experiments proved successful.

Later, with a few employees, a water-wheel powered lathe, and a lot of determination, the two young industrialists went to work. Their course was quite clear then, as Italy produced virtually no motorcycles and her export business was non-existent. In this atmosphere, one would think that chief engineer Guzzi would have stuck to orthodox or proven design principles, but this was not the case, and his approach was to be a really fresh one to the problem.

In the December 15th issue of *Motociclismo,* the new machine was announced publicly, and the name changed to Moto Guzzi. The specifications were rather advanced for those early days, and also very different, but on this design, the infant company was to stake its future. The frame and fork were quite standard for those early days, but the single cylinder engine was mounted horizontally in the frame. Not only was that unusual, for the flywheel was mounted outside of the crankcase instead of inside.

The valve gear was the inlet-over-exhaust arrangement, which meant that it had an "overhead" inlet valve and a "side" exhaust valve. The engine capacity was 500cc, and a hand shift and chain drive were featured. The front wheel had no brake and the rear wheel used an internal expanding brake.

Now, one of the principles that has historically been associated with Guzzi has been that racing improves the breed, and so this "La Prima" model received an early baptism. Aldo Finzi is the name that the history books mention with a win at the famous Targo Florio circuit, but it was really the amazing stamina of this new machine that impressed the public.

Not content to sit on his laurels, Carlo kept at his design work, and in 1923, he produced a new overhead valve engine, as well as a production racing machine to go with the standard road model. This new 500cc model also featured the horizontal engine with the tiny crankcase and outside flywheel and it, too, was soon entered in the great homeland races to prove the brilliance of its design. And it won, taking trophies at Concorso Di Consumo, Giro d' Italia, Pinerolo-Costagrande and Del Lario.

These early successes went a long way in proving the genius of this young Ing. Guzzi, and public demand for his products rapidly increased. Once again, new designs came off the drawing boards, and this time a 250cc single joined its 500cc older brother. The new engines featured a bevel-gear-driven overhead cam, and four valves were used along with two exhaust pipes. The girder fork was retained on the front and an internal expanding brake was still used on only the rear wheel. The frame was the old "vintage era" type with the gas tank between two top frame tubes and the oil tank mounted on top of that.

These 1924 models were raced with great success in Italy. Then, in September, the Guzzis were prepared for their first international race — the Grand Prix of Europe at Monza. Despite the really fierce opposition from all the leading factories in Europe, the marque scored a smashing victory. Guido Mentasi took first, E. Visioli garnered second and Pietro Ghersi captured third — a truly remarkable achievement for such a tiny factory.

This great international victory put the company firmly on the map and sales rapidly expanded. For 1925, a new "Peoples Sport" model was introduced which had the I.O.E. type of engine, lights, and a three-speed hand gearshift.

A few years later, in 1928, the engineering progress was really manifest when the 500cc "Sport 14" and "Grand Turismo" models were announced. The engines were a new ohv design with a hemispherical combustion chamber, and much larger fins were used to dissipate the heat. A dynamo was fitted for the lighting system and a magneto was used for ignition. Lastly, internal expanding brakes were finally fitted to both wheels, which made the bike much safer to ride.

A definite attempt was also made to improve rider comfort as a more modern girder fork with both compression and rebound springs was used. It was the Grand Turismo model that really made the news for rider comfort, as it included a method of rear suspension in its specifications. The springing was effected by having a triangulated rear sub-frame which was pivoted at its top joint to the main frame and with the lower arm connected to a long spring-box that ran completely underneath the engine. This was probably the first really successful spring frame on a motorcycle, and it was truly a great milestone in engineering progress.

The following year, the factory again made great strides in producing a good machine for the public. Use of the motor-

Keith Campbell, on his way to victory in the 1957 race on the 350cc single. The 350 model weighed only 216 pounds and the 500 only 235. Top speeds were about 145 and 160 mph.

cycle for both sport and transportation was on the increase, and the public demanded a more reliable mount that was capable of genuine touring. Guzzi responded with a new range of 250 and 500cc models for 1929, and one of the most notable changes was the new triangulated cradle frame that had the "saddle" type gas tank mounted over the top frame tube with the oil tank beneath the gas tank.

That same year Carlo Guzzi manufactured his first motorcycle-truck combination – a type of machine that is still produced today. The idea was sound, since Italy was still a rather poor country, and this inexpensive method of hauling materials was just the ticket for the businessman.

The furious design improvement pace was maintained and during the early 1930s, many new models were introduced. The 1931 "Sport 15" model proved to be particularly popular; its clean lines were a big improvement over the earlier models. By 1934, a customer had his choice of a 175, 250 or 500cc Guzzi in standard trim, with a spring frame, or to road racing specifications.

The 500cc production racing model achieved many successes during the late twenties and early thirties in continental grand prix events and national championships, but the marque had failed to score any really big victories that would enhance export sales. The traditional oversquare ohc 500cc single with measurements of 88 x 82mm bore and stroke was pumping out only about 97 mph in 1930, so the factory decided to put more emphasis on some new designs that would set the racing world on fire.

In 1933, the work began on a new 500cc twin that was destined to usher in an exciting new era for the company, as well as take them to the summit of motorcycle racing – the Isle of Man TT. It was not until 1935 that they actually got there, and that year happened to fall in the middle of the era when anyone with any sense at all left the all-conquering Norton strictly unchallenged.

Norton had a famous Scot by the name of Jim Guthrie, who piloted his singles with such skill that he simply squashed the opposition into embarrassing second places. Realizing full well that it would take a truly great rider, plus a mighty fast Guzzi, to upset this reigning world champion, the factory hired one Stanley Woods to do the coup de grace. Stanley was, in his own right, a formidable contender, having won more races and TTs than anyone else, but to trounce the Norton on its home ground was asking quite a bit.

The new Guzzi, though, was quite a racing machine. The basic design was a wide angle (120-degree) V-twin, which was rumored to have offset cranks so that it fired at 180 degrees. Bevel gears and shafts drove the single overhead cams, and hairpin valvesprings were used. The engine produced about 44 hp at 7,500 rpm for a maximum of 112 mph, which was about equal to the Norton single.

The front fork was an orthodox girder type, but the rear wheel had a very unusual feature for the race track – a type of pivoted fork suspension. The fork was a triangulated affair similar to that used on the 1928 GT model, but the spring boxes were mounted beneath the rear fork, so that the spring unit moved with the fork. A springer had never won a Senior TT, and it was generally agreed that the handling would never be good enough and the whole plot would be unstable at speed. This suspension, incidentally, could have its damping altered during the race by simply moving a lever on the left side of the bike.

The other specs on the twin were equally impressive. A massive brake was fitted to the front wheel which had twin leading shoes, a four-speed footshift gearbox was used, and straight pipes were used instead of megaphones. Another innovation was alloy wheelrims – a move to cut down unsprung weight as well as lessen the revolving weight that the brakes must stop.

And so it was. The race began as everyone thought it would, with Guthrie leading Woods by 28 seconds. On the second lap the margin was 47 seconds, and on the third lap it was 52. Then, on the fourth lap, Stanley began to creep up and the clocks showed only 42 seconds difference. On the fifth it was 29 seconds, and on the sixth it was 26. Still, 26 seconds was a very substantial margin to make up on such a brilliant rider as Guthrie, so it looked like another Norton win.

At the end of the seventh lap, Guthrie wheeled in to be acclaimed the winner, the wreath was placed around his neck, champagne was uncorked, and the Island's Governor gave Jim his congratulations. Guthrie, by virtue of his starting number one, was the first to finish, while Woods was well back in the pack on a much later starting interval. So, all there was for Jim to do was to watch Stanley's clock and see just how bad he had him beat.

Then began the drama. Into the last lap went Stanley, his torso glued to the top of the tank and the V-twin turned up to a noticeably higher speed. Faster and faster he went, and the revs mounted to 8,000. From curb to gutter the Guzzi was flung

The fabulous 1957 Guzzi V-8 is shown without its streamlined fairing. Despite the bulk, weight was only 320 pounds. Horsepower was about 75 at 12,000 rpm.

The "La Prima" Moto Guzzi — 1921 model. The engine is an inlet-over-exhaust-valve type and it proved to be very reliable for its day.

Seldom seen and never raced was the 1940 Guzzi 500cc "three." The war halted development of this potent looking mount of which little is known. Double overhead cams appear to be chain driven.

A cobby looking racer was the 1935 wide angle twin which won the Senior TT at record speed. Horsepower was 44, top speed about 112 mph.

The 1952 Gambalunghino 250cc model was the father of the 350cc models that won World Championships in 1953, '54, '55, '56 and 1957. Model featured a five-speed gearbox, 27 hp.

and the seconds were slowly pegged back – one by one. Silently in his pits, Jim Guthrie watched as Stanley and the Guzzi gained. Soon he came into sight and screamed past the checkered flag.

It took a few minutes to calculate, but then came the announcement that Stanley had smashed the lap record by nearly four miles per hour on his last lap and so had won the Senior TT by four tiny seconds. Overnight the marque became famous, as Stanley had also won the Lightweight TT on his little single. A new era was born.

Back in Italy, the factory expanded its plant facilities to keep pace with the demand for their products, and soon the roadsters were featuring such race-proven features as footshifts, powerful brakes, speedy ohv engines and superb roadholding.

In the race shop the midnight oil was burned and many brilliant designs were produced. Probably the most successful was the ohc 250cc single which won many races, as well as setting a speed mark of 132.2 mph in the hands of R. Alberiti, and a standing start kilometer record of 25.45 sec. (88.18 mph average) by G. Sandri. This little 250 featured a supercharger, which is one of the few blown singles to ever be successful, and it even scored an astounding first and second place in the 1939 German Grand Prix when Nello Pagani and Sandri trounced the reigning world champion DKWs.

Another blown Guzzi was the 1940 three-cylinder model, a 500cc beast that never saw many races because of World War II. Probably the prettiest of the lot was the 1939 Condor, a 500cc production racer that established a splendid reputation in the hands of private owners. The Condor featured the pivoted-fork rear suspension pioneered on the 1928 GT model – one of the first uses of a spring frame on a production racer.

Then came the war and all motorcycle racing and most motorcycle production was halted in Italy. After the war, the factory was soon into production; many new street models were produced in sizes ranging from 65cc on up to 500cc. Scooters, motorcycle-trucks, 250cc roadsters, 500cc roadsters, and production road racers were all manufactured, and the marque's sales hit a new high as exports spread all over the world.

Probably the most popular Guzzi was the 250cc Airone Sport, a race-bred model that provided a spirited 80 mph top speed combined with powerful brakes and a good suspension. The Airone featured an ohv engine with bore and stroke of 70 x 64mm, and the powerplant acquired a reputation of being unburstable.

The 500cc models were the Astore (touring) and the Falcone (a sports model), and these Guzzis were known for their excellent roadholding and tireless high cruising speeds. The Falcone was clocked at 98 mph in European road tests, but few ever made their way to America.

In the late 1950s, the horizontal singles were phased out and new, more modern looking vertical-mounted engines made their debut. These new scooters and bikes up to 250cc have maintained a healthy sales figure and have continued the Guzzi reputation of quality engineering. Realizing that perhaps something was lost when the big singles were dropped, the factory recently began production of a 700cc V-twin, which was originally intended for police use.

The marque also resumed racing activities after the war and their successes reached a new high. A new 250, the Albatross, made its debut in 1945, and the model was also sold as a production racer. The Albatross was a cobby looking little bike with its traditional ohc horizontal engine, unusual pivoted-fork rear suspension, and the girder front fork. The engine turned to 8,000 revs and produced 24 hp, good enough for a speed of 102 mph.

The 250's bigger brother was the Dondolino, a 500cc model that also was available to the private owner. A really great classic of a racing bike, the big single never did achieve the aura of invincibility that surrounded the Albatross. The 250 literally dominated its class in 1947 and 1948, with the 500 scoring a few victories such as the 1947 Swiss and 1948 Ulster classics.

For 1949, the marque once again introduced new singles, both as works and production racers. The new 250 was called the Gambalunghino and the 500 was named the Gambalunga, which meant simply "little long leg" and "long leg." These two bikes continued the Guzzi tradition of winning races. Also raced in those early post-war days was the old V-twin, which had undergone a great deal of development. The twin's power was down to something like 42 at 6,800 rpm, though, as the post-war switch from petrol-benzol fuel to just plain "pool" gas forced a drastic lowering of compression ratios.

The twin did not fare too well in those days, as the reliability was not up to the standard of the singles. In the 1948 Senior TT, for instance, Omobono Tenni had a big lead at the end of the fourth lap until one cylinder failed and he toured home in ninth position. Then, in 1949, Bob Foster was way out in front with only a lap and a half to go when the clutch gave up. The development work went on, however, and in 1952, the engine was producing 48 hp at 8,000 rpm for a top speed of 129 mph. The twin did gain many good "places" in those years, but the factory decided that the one win in the 1951 Swiss GP was not enough to keep the model in the lineup.

The little 250 single did much better, and its reputation as a winner was sealed with such victories as the 1949, '51, '52 and '53 Lightweight TTs, the World Championship in 1949 (Bruno Ruffo), 1951 (Ruffo), and 1952 (Enrico Lorenzetti), and many hundreds of minor grand prix events. The Gambalunghino, by 1952, was producing 27 hp at 7,500 rpm on its 8.7 to 1 compression ratio, and this was good enough for a 109 mph top end. The little racer was noted for its fine handling, its unique feature of the gas tank being streamlined for the rider's arms, the leading link front fork, a massive front brake, and the horizontal engine with its outside flywheel.

About that time, the leading designer at the factory, Ing. Giulio Carcano, decided to join the Italian trend to four cylinders – only with a difference – as the Guzzi four had the engine mounted in line with the frame. Shaft drive and water-cooling were adopted, as was a partial "dolphin" type of fairing. The new four had a very advanced dohc head with fuel injection, and four short straight pipes were used. The four was very fast and it won its first race at Hockenheim in 1953 when Lorenzetti romped home at an average speed of 107.8 mph. After that initial success the four turned out to be totally unreliable and it was soon dropped.

A far more successful chapter was the quest of the 1953 World Championship in the Junior (350cc) Class. Fergus Anderson was on the team then, and, after his record breaking rides in the 1952 and 1953 Lightweight TTs, he urged Carcano to scale up his 250 into a 350. The first step was to increase the 68 x 68mm en-

Continued on page **58**

The 1953 four is shown on its record breaking ride in the Hockenheim classic. Note the leading link front fork and shaft drive. The four proved to be unreliable.

The 1947 Albatross was a very successful 250cc racer in the post-war years. Note the spring box for the rear suspension beneath engine. Engine produced 24 hp at 8,000 rpm.

MOTO GUZZI

gine to a 72 x 80mm size, which gave 320cc. This move was eminently successful and Fergus garnered a third in the Junior TT and then a win at Hockenheim.

Justifiably encouraged, the factory enlarged the engine in mid-season to a 75 x 79mm 350, and power was thus raised from 28.5 at 8,400 rpm to 31 at 7,700. The weight of this Junior model was a very light 264 pounds, and the maximum speed was an amazing 130 mph. This terrific speed on only 31 hp was due to another great Guzzi engineering achievement – streamlining. The racers that year showed the results of many hours spent in the wind tunnel on airflow studies, and the full "dustbin" fairing gave excellent air penetration. On this speedy 350, Fergus Anderson easily won the World Championship.

For 1954, the valve sizes were increased to 37mm for the inlet and 32mm for the exhaust, a 35mm carburetor was fitted, and the compression ratio was 9.5 to 1. Power output was 33.5 at 7,500 rpm. Anderson was again successful and one more World Championship came home to Mandello del Lario.

In 1955, the brilliance of Ing. Carcano was once again manifest when the racers appeared with a new and very light "space" frame that was designed to fit the dustbin fairing. The engine was also modified to a 80mm bore and a 69.5 stroke, the valves enlarged to 41 and 36mm, and the carburetor increased to 37mm. The head featured two spark plugs, battery ignition replaced the magneto, and the compression ratio was 9.4 to 1. And then, lastly, the head was modified to a double overhead cam setup; all these improvements helped boost power to 35 at 7,800 rpm.

Bill Lomas was the first-string rider for 1955 and 1956 and he easily won the 350cc world titles, but on the tighter courses the new engine had actually lapped slower than the old long stroke engine. Ing. Carcano once again made some changes for 1957 when he went back to the 75 x 79mm measurements with 39 and 33mm valve sizes. One spark plug was used and ignition was by a magneto. Power output was up to 38 at 8,000 rpm and the medium speed torque was much improved. The weight of these racers was fantastically light, with the 350 weighing only 216 pounds and a larger 500 model just 20 pounds more.

Carcano's genius, however, did not end here, for he was hard at work designing the fastest road racing motorcycle that the world has ever seen. First raced in 1956, it was not until 1957 that the world found out just how fast this new 500 was. The basic design was a dohc 90-degree V-8 with the tiny measurements of 44 x 41mm bore and stroke. Eight 20mm carburetors were used as were eight short exhaust pipes, and water-cooling was also employed.

The engine was mounted transversely in the frame and chain drive was used. The frame was similar to that used since 1953 on all Guzzi racers, with a swinging-arm rear suspension and a leading link front fork. Massive brakes were fitted to cope with the colossal speed. The weight was actually quite light at 320 pounds, but the sheer bulk of the bike made it appear much heavier.

It is only natural that many teething troubles cropped up on such a complex machine, and lap records, rather than race records, fell to the V-8. The horsepower in 1956 was listed as 62 at 12,000 rpm, which provided a 162 mph top end on a 6.1 to 1 gear ratio. This speed, however, was with the dolphin-type of streamlining, and it was not until one year later that the fabulous eight was to truly show its speed.

The venue was the beautiful Spa-Francorchamps course in Belgium, which is terribly fast with its 100 mph corners and long straights. Keith Campbell, from Australia, was the first-string rider that year, and his skill was never in doubt, as he had already trounced the best in the 350cc class. But would the new V-8 hold together long enough to actually win a race?

The flag fell and away went the field with such stars as Libero Liberati on his Gilera, John Surtees, MV Agusta, and Walter Zeller, BMW. It was soon apparent that the 75 hp the V-8 developed, combined with the full dustbin fairing, gave Campbell the speed he needed to sear through the field and take the lead. On and on Keith rode, smashing the lap record again and again, until he turned in his finest hour with a 118.56 mph lap. Then, just after he was clocked at a fantastic 178 mph on the fast Masta straight, Campbell came to a halt with gearbox trouble.

And so ended the last great chapter of the fabulous Moto Guzzi V-8, for little did the half-million spectators know that an era was ended. For at the season's end, the beloved racing Guzzis, along with their World Champion brothers, the 125 and 250cc Mondial and 500cc Gilera, announced that they were withdrawing from racing. It seemed impossible at first that Guzzi could really mean this, for their heritage of racing since 1921 was one of the greatest, and certainly among the most colorful, that has ever existed.

We realize now, of course, that the real cause for their decision was not within their control. As Italy prospered in the post-war economy, her people switched from motorbikes to cars, and this does not help a motorcycle company's profits any. It takes great sales and huge profits to design a modern-day racing bike and then take it all over the world to do battle.

Sadly, it all came to an end. The men, the great races, and the romantic machines – it is all just a legend now. But while it lived, it was one of the really fascinating stories of our century, and the uninhibited engineering and beautifully exotic racing bikes will probably never be equaled.

From 1921 through 1957 –a truly remarkable record. For their honor, I list it below:

International Racing Victories	3,329
World Speed Records	134
Italian Championships	47
National Championships	55
TT Victories	11
World Championships	14

■

CYCLE WORLD
ROAD TEST

MOTO GUZZI V7

THE MOTO GUZZI V-7 is exceptionally quiet — on the road, that is. This particular motorcycle starts to generate noise only at the end of the trip, as soon as the ignition key is flipped to the "Off" position and the side-stand makes contact with driveway concrete.

"What's the horsepower?"

"Does it run as smoothly as a BMW?"

"Do your shins get hot from those cylinders up there?"

Everywhere the Moto Guzzi V-7 makes an appearance, it attracts a crowd, and members of the crowd ask questions. Enthusiasts who never before have seen this awesome Italian in red, black and silver rococo splendor arrive on creepers from under cars, at a dog-trot from across streets, and on quickly-braked motorcycles of other manufacture.

Questions seem to arise naturally at first view of this huge, magnificent touring twin. Will its 90-degree V engine layout, incorporating minor design changes for 1968,

prove itself a shaker with adverse torque troubles? Will absence of a kick lever in favor of an electric starting system prove difficult? What about brakes to accommodate what appears to be a very gross gross? Will sheer size and weight offer handling problems? So much for questions. With the Moto Guzzi V-7, discovery of answers is far better sport.

From the outside, the Moto Guzzi 703.7-cc/50-bhp V-twin engine is nothing short of impressive. Castings and turnings exude quality, the essence of which is craftsmanship; absence of leaking lubricants indicates care in design and fabrication. Deep crankcase, finned oil sump, and massive cylinder barrels, broad enough to completely encompass pushrod tubes with their fins, create a distinct impression that Moto Guzzi has built this machine almost entirely of engine.

Inside this huge powerplant's crankcase, a single-journal crankshaft turns in two main bearings. This crank is heavily counterweighted to damp vibration effects of two-cylinder, four-cycle firing eccentricities. The left-hand cylinder bore center is slightly forward of its right-hand counterpart. Thus it follows that the big end of the left-hand connecting rod rides in the single journal ahead of the right-hand rod big end.

The camshaft, which turns in the point of the V directly above the crank, and the oil pump are helical gear driven off the forward end of the shaft. The center gear of a three-gear tower is the crankshaft pinion; the bottom gear drives the oil pump; and the top gear drives the camshaft. The spark distributor also is spiral bevel gear driven, but from the rear of the camshaft.

The pushrods actuate the valves through rocker arms. Moto Guzzi employs single, rather than compound, valve springs in the V-7 engine, something of a departure from current motorcycle and automobile practice.

A 12-V/32-ampere DC generator, located between the cylinder barrels, is driven off a very automotive V-belt and pulley system at the nose of the crankshaft. The electric starter, key-switch operated, located at the left side of the engine, drives through Bendix-type ratchet engagement to the toothed flywheel rim.

Carburetion for the V-7 is through a pair of Dellorto units of 29-mm venturi diameter. These appeared to run slightly lean during testing, a condition that could be easily rectified by an exchange of jets.

Power is delivered from the engine to the four-speed gearbox through a two-plate dry clutch, another component of the V-7 which bears great resemblance to part of an automobile. The drum-shift, two-shaft transmission itself, however, is pure motorcycle in design and operation. As the right foot rocker lever rotates the drum through a system of pushrods and bellcranks, the selector forks move the gear shaft forward or backward for engagement of one or the other of four constant mesh ratios.

Output from the gearbox is through a double universal joint to a splined cardan shaft, which runs in ball bearings forward, and tapered roller bearings at the rear, and through a spiral beveled pinion and ring gear set in the rear wheel hub. Splined at either end, the cardan shaft functions as a slip joint to accommodate movement of the rear swinging arm suspension. The right suspension arm is the driveshaft housing, and is bolted rigidly to the rear hub casing. The drive axle assembly turns in one ball-type thrust bearing to the right, and a massive load-carrying ball bearing to the left. The expense that goes into manufacture of shaft drive layout, in the case of the Moto Guzzi V-7, is well rewarded in smooth power delivery, and silent operation that no chain and sprocket system can provide.

Action of the rear suspension is hydraulically damped by a pair of telescopic shock absorbers and exposed coil springs. Three-position adjustment provides for solo, medium, and heavy load riding situations. (Tire pressure is critical in setting up the rear suspension for two-up touring. Though the manufacturer recommends 21 psi for the front tire for all forms of operation, recommended rear tire inflation is 25 psi for solo riding, and 28 psi with a pillion partner. Experiments showed greatly improved handling when suspension and tire adjustments were made in accordance with the manufacturer's recommendations.)

The telescopic front fork of the V-7 is greatly reminiscent of certain products from Milwaukee, Wis. Massive steel upper and lower fork crowns support and maintain alignment of the telescopic fork tubes. As one observer said, "You couldn't bend 'em if you ran over 'em with a truck."

Heavy-duty describes the Moto-Guzzi V-7's Duplex frame. Large diameter tubing, muscular forged joints, unusually huge rear arm pivot, and exceptional thickness of plate materials indicate super strength, durability, and, of course, longevity. Moto Guzzi V-7 solidity of construction invites long-distance, high-speed touring.

Though the hefty Italian machine is no sprinter, its accelerative ability is creditable enough in view of total weight comprised of machine, fuel and rider. Top speed far exceeds the speed limit of any of the 50 states, save Nevada.

What goes this well must be made to stop with equal efficiency. Large diameter drums carry double leading shoes forward and a single leading shoe at the rear. In the test machine, the rear shoe appeared to be not at all well seated and protested with a groan when used; thus the major portion of deceleration work was left to the front system. Little hand lever effort was required for an all-on stop (with the rear brake used lightly, only for directional stability). Though the Moto Guzzi seemed to rush to a stop, negligible forward dive was encountered. To sum up, the V-7 is a big machine, which has big brakes, which do a big job very well.

Rider/passenger accommodation of the V-7 is equal to, perhaps surpassed by, only one or two machines in the world — and these points of superiority are debatable at length and certainly subject to individual rider preference.

Fat, spongy, Italian-style twist grip handles absorb the numbing vibration of a long, high-speed run. Lighting and choke controls are within easy thumb's reach, left and right, respectively. The rocker-type shift lever (a device almost entirely peculiar to Italian two-wheeled machinery) requires practice, but once accustomed to toe for up shifts, heel for down shifts, riders start to express a preference for this system. For the passenger, in addition to the conventional grab strap across the seat, Moto Guzzi fits a pair of chromed tubular steel handles, one at either side of the saddle.

The V-7 arrives from the factory with a pair of bars forward which are designed to protect both the protruding cylinder barrels and the rider's legs in event the machine tips over. Whether these bars would withstand a crash of any great severity is a matter for speculation, but not test.

Twin cam/plunger fuel cocks are located forward on either side of the tank. These valves are neither given reference in the V-7 *Rider's Handbook,* nor marked for "On" or "Off." Both must be in the off position when the machine is at rest, else the float bowls will eventually overflow and drizzle gasoline onto the crankcase and surrounding areas.

Quibbles with small irritants can in no way minimize the outsized pleasure of riding this immense road machine. Half choke, fuel valves on, ignition switch on, neutral indicator light glowing yellow — then the key is turned full right. Two or three revolutions from the 0.7-bhp starter motor and the V-twin engine comes to life with the bass purr of a well-fed tiger. Heel down puts the drive train in first gear. Slight application of throttle and easy release of clutch results in one of the smoothest starts from rest in all of motorcycling.

Stops and starts through the red-light-green-light, stop and go metropolitan traffic reveal the balance, low speed maneuverability and docile nature of the V-7. On the way to the country, the big Moto Guzzi demonstrates its fast cruise at 65 mph; and its freeway manners are impeccable — effortless lane changes, forthright spurts of speed for safe passing of other vehicles, and a feeling of security elicited by few other motorcycles.

Though the Moto Guzzi V-7 is eminently suitable for cross-town junketing and fast expressway tours, the delightful roadability of this machine comes to the forefront on the rural two-laners where every bend is a challenge, where the sweeping curve is met with an inclination of the head to adjust its radius, and where the tight S-bend is simply a sure whip-whip action.

Touring two-up with the V-7, machine and passengers total some 800 lb. in weight. Extensive test riding proves the only place the 700-plus-cc engine is likely to bog a bit with two passengers is at 70 mph, in top gear, up a long, straight incline, into the teeth of a 20-knot headwind. A quick drop into third gear remedies the problem with added revolutions and added torque.

The Moto Guzzi V-7 is priced at $1500. This is something more than the $1300 asked for 600- and 650-cc Japanese, German and British machines. However, the extra $200 in price for the V-7 does bring the electric starting feature, an added 50-cc piston displacement, shaft drive as with BMW and Marusho, and that immense, crowd-gathering conversation piece of a transverse V-twin. The latter is worth at least $200 in making new acquaintances who wish to talk about Italian vs. German, or Italian vs. English, or Italian vs. Japanese machinery.

The dyed-in-the wool touring motorcyclist will find the Moto Guzzi V-7 a delight to own. His motorcycling cronies will find it an everlasting source of envy. The Moto Guzzi is one of a very small number of motorcycles that can create a great deal of noise, while remaining very smooth, very quiet. ■

700-cc MOTO GUZZI V-7

SPECIFICATIONS

List price	$1454
Suspension, front	telescopic fork
Suspension, rear	swing arm
Tire, front	4.00-18
Tire, rear	4.00-18
Brake, front, diameter x width, in.	8.66 x 1.66
Brake, rear, diameter x width, in.	8.66 x 1.66
Total brake swept area, sq.-in.	90.1
Brake loading (test weight/swept area) lb./sq.-in.	7.34
Engine type	4-cycle, ohv 90° V-twin
Bore and stroke (inches-millimeters)	3.149 x 2.775, 80 x 70
Displacement (inches^3-centimeters^3)	42.9, 703.7
Compression ratio	9:1
Carburetion	(2) 29mm Dellorto SS1
Ignition	battery and coil
Bhp @ rpm	50 @ 6000
Oil system	pressure, by gear pump
Oil capacity, pts.	6.15
Fuel capacity, gal.	7.57
Starting system	electric motor
Lighting system	generator and battery (12V)
Air filtration	paper element
Clutch	twin driven dry plates
Primary drive	gear, 1.352:1
Final drive	double joint cardan shaft, bevel gears, 4.625:1
Gear ratios, overall:1	
5th	none
4th	4.564
3rd	6.077
2nd	7.816
1st	11.330
Wheelbase	57.5
Seat height	31.5
Seat width	12.2
Foot-peg height	12.1
Ground clearance	6.0
Curb weight (w/half-tank fuel)	502
Test weight (fuel and rider)	662

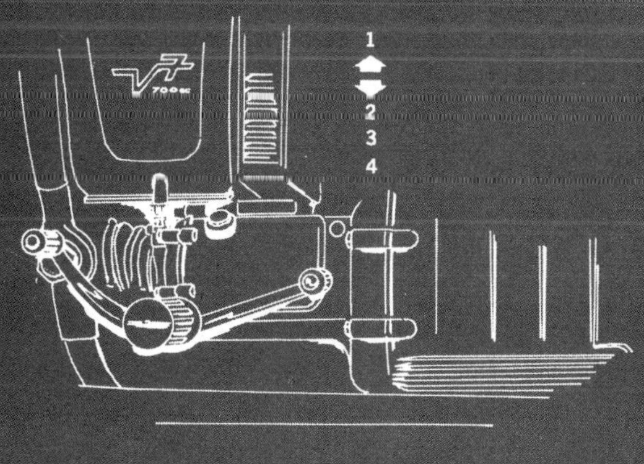

PERFORMANCE

Top speed	98
Maximum speed in gears (@ 6000 rpm)	
5th	none
4th	98
3rd	74
2nd	57
1st	39
Mph per 1000 rpm, top gear	16.3
Speedometer error	
30 mph indicated, actually	28.6
50	49.2
70	68.4
Acceleration, zero to —	
30 mph, sec.	3.1
40	4.6
50	6.8
60	9.3
70	12.1
80	16.1
90	21.9
100	
Standing 1/8-mile, sec.	10.40
terminal speed	65.50
Standing 1/4-mile, sec.	16.31
terminal speed	81.15

ACCELERATION AND ENGINE / ROAD SPEED
RPM X 100

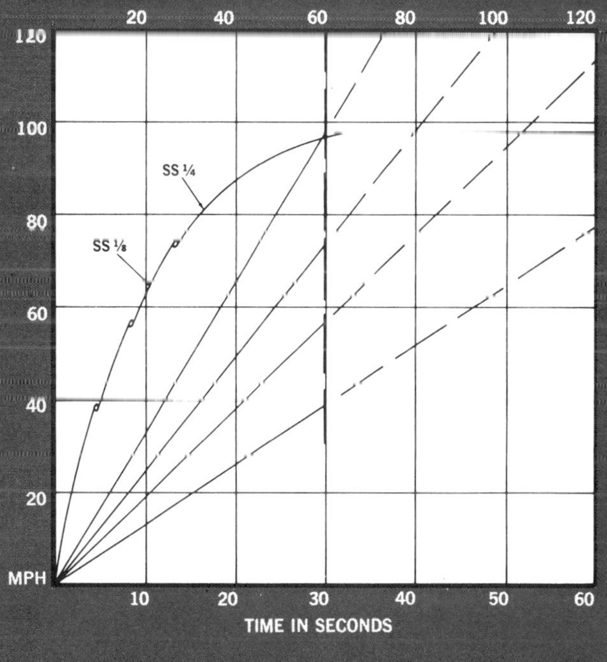

Nearside showing the electric starter at the rear of the heavily-finned sump. The position of the V-twin cylinders causes the rider to sit well back on the machine. The twin carbs turn through almost 90 degrees to air filter

A Marelli distributor and coil mounted between the cylinders supply the sparks. A heat gasket is fitted between the induction stub and the cylinder head (note Allen screws). The push-rod operated valve gear is below polished covers

Right hand side of engine shows the rocking pedal gearchange and heavy duplex frame. The crash bars are fitted as standard. The 300W(!) dynamo, mounted between the cylinders, is car-like and belt-driven from a crankshaft pulley

MM SUPER TEST

GIANT

Look at some motorcycles and you can immediately sum up their personality . . . but not the 700 cc Moto Guzzi V7. It looks clumsy when, in fact, it handles precisely. It appears supremely powerful, but won't top the ton. In short, the Guzzi V7 is a machine full of surprises—none of them really unpleasant.

The giant V7 derives its name from the almost unique "across the frame", opposed vee-twin engine layout. Other manufacturers have tried this arrangement without a great deal of success. The Lambretta scooter concern had an opposed vee-twin racer that appeared in a few 250 cc races in the early 1950s, the German Victoria company had a 350 cc version which they called the "Bermeister" or "Mountain Climber".

This never climbed very far up the marketing mountain and the Lambretta didn't exactly make Count Agusta's hair fall out, so neither of them could really be termed as highly-successful.

However, the Guzzi, V7 does exactly what its makers intended. They decided on a big, comfortable touring machine, with no pretensions to racing speeds or handling and that's exactly what they built.

V-formation

The engine is actually 703.7 cc with a bore and stroke of 80 mm. x 70 mm. push-rod, ohv four-stroke, its cylinders are opposed in a V-formation, with the angle between the cylinders being 90 degrees.

Set across the frame, similar to the BMW, it drives through a two-plate, car-type clutch, via a four-speed gearbox and a shaft final drive to the rear wheel. Using a car-type layout gives car-type smoothness and although there is slight engine vibration under heavy load at low revs, there is virtually no vibration throughout the rest of the 6,000 rpm range of the big twin.

With 700 cc poking out 50 brake-horse-power, it's surprising that the Guzzi V7 will not top the ton. But the revs and power seem to be used not for top end speed so much as tractability and accelerating the 500 plus pounds at a reasonable rate—the standing quarter was covered in just over 17 sec.

Top speed is actually a shade over 98 mph, which is plenty fast enough for the touring rider, especially with the Government still insisting that 70 mph is quite fast enough for anyone . . . man, woman or beast.

Tractability

One can't help comparing the Guzzi big twin with the BMW. It has the same loping tractability which allows a purring 20 or 80 mph plus in top gear. There is none of the screaming, roaring exhaust note of the lightweight and you really need an effective horn to let straying pedestrians know you're coming.

Throughout the test, the engine and transmission remained completely oil-tight and it was simply a matter of a quick wash down to remove road dust and dirt to bring the machine back to sparkling showroom finish. The fully-enclosed, shaft drive added even more to the ease of this normally miserable chore.

I have already stated that the Guzzi V7 was obviously designed for the long-distance rider and everything possible is included in the machine's specifications to provide minimum effort and maximum comfort for the owner.

An electric starter eliminates the possibility of kick-start ruptures and the designers have kept the lavishly padded dual seat down to a height of 31 inches, which means that only those with the stubbiest legs are likely to have trouble footing the 500 lb. monster through the hurly-burly of town traffic jams.

Precise steering

Though its huge, alloy cylinders protrude from an equally massive, finned crankcase combined sump, making the Guzzi look somewhat clumsy, this is in fact far from the truth. The handling and steering are precise enough for normal touring purposes, although the sheer bulk and weight of the machine means that it can't be flicked through tight, country lane bends similar to a lightweight.

The handling and steering is akin to the big American Harley and the bike is obviously designed to appeal tc this market.

To carry a load of 500 lb., plus two passengers—making a combined weight of 800 lb. or more, the suspension has to be more than robust. For this purpose, the Guzzi is equipped with monstrous telescopic forks and rear swinging-arm units. The damping of these units is really effective and like a big car, the machine rides smoothly on the roughest road surfaces.

To cope with varying loads, the rear suspension legs are adjustable to three positions for light, medium and heavy. This, combined with equally large 4.00 x 18 tyres on front and rear wheels, gives an armchair smooth gallop, making long

he bevel gear housing of the shaft drive on the iuzzi supports the suspension leg, while the haft drive outer casing acts as one half of he swinging arm. Note the wheel rims are all lloy Borrani units carrying 4.00 x 18 in. tyres

The rear $8\frac{1}{2}$ in. single-leading shoe brake is exceptionally clean in design, making for very easy polishing. The suspension legs are adjustable for load, providing a smooth ride under all road conditions. Note pinch bolt on q.d. wheel

This is the real stopper. In spite of lack of cooling air scoops, the $8\frac{1}{2}$ in. twin-leading shoe front brake is really effective at any speed. The massive fork legs are of Guzzi design and have two-way damping to give superb steering

GUZZI

REPORTING FROM THE STATES—BRUCE COX TESTS AND DESCRIBES THE UNIQUE 700 cc V7

MOTO GUZZI V7

MM SUPER TEST

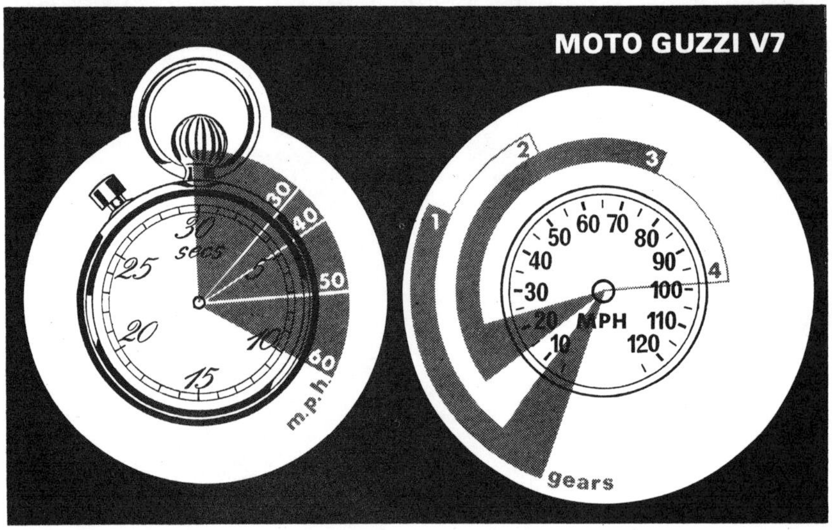

Performance:

Speeds in gears:

(in mph)	Minimum	Maximum
Bottom	—	40
2nd	8	55
3rd	16	75
Top	23	99

Acceleration:

0–30 in 3.5 sec. 0–50 in 7.2 sec.
0–40 in 4.8 sec. 0–60 in 10.0 sec.

Standing start quarter-mile covered in 17.0 sec. with a terminal speed of 81 mph.

Fuel Consumption:
Average on test—47 mpg.
Touring—52 mpg.

Brakes:
Braking from 30 mph—34 ft. 2 in. on dry tarmac with 12½-stone rider.

Specification:

Engine details:
Ohv 90 degree twin. Bore and stroke 80 × 70 mm. giving 704 cc. Compression ratio is 9 : 1 giving a claimed 50 bhp at 6,000 rpm. Light alloy barrels with chrome bores. Steel con-rods with plain shell big-ends. Wet sump lubrication with gear-driven pump.

Transmission:
Primary drive by gear. Four-speed, foot-controlled gearchange, constant mesh. Two-plate dry clutch fitted with "cush" shock absorber. Final drive by Cardan shaft, double-jointed to bevel gear at rear wheel.
Gear ratios; 1st—11.3 2nd—7.8
3rd—6.0 top—4.6

Electrical Equipment:
12-volt distributor and coil. 300W dynamo. Electric starter. 12v. 32ah. battery.

Carburation:
Twin 29 mm. Dell'orto SSI carburettors with paper element air cleaners.

Wheels:
Front—4.00 × 18 tyre carried on alloy rim. 8½ in. twin-leading shoe front brake.
Rear—4.00 × 18 tyre carried on alloy rim. 8½ in. single-leading-shoe, rod-operated brake.

Dimensions:
Fuel tank capacity—4½ gallons.
Wheelbase—4 ft. 10 in.
Seat height—2 ft. 7 in.
Ground clearance—6 in.
Weight—502 lb.
Manufacturers: Moto-Guzzi spa, Mandelio del Lario, Como, Italy.
Probable price—£600.

GIANT GUZZI V7

distance touring completely effortless.

Stopping an 800 lb. mass of machinery and passengers from 80 mph requires better than average brakes and, once again, the Guzzi designers have studied the problem and produced a fair solution.

A total brake area of 90 square inches is covered by the double-leading shoe front and single-leading shoe rear brakes, which are housed in 8½ in. drums. Very little effort is required to bring the machine to a tyre-squealing halt and there is surprisingly little front end dip from the suspension, even under crash stop tests.

Electrics

The electrics on the Guzzi are 12-volt and with the lighting system designed to pass stringent American regulations, both front and rear illumination is equal to any of the machines tested by MM.

Ignition is by battery and coil and with a large capacity battery to power the electric starter, whirling the twin into life is simplicity itself. With the ignition on, choke half closed and the neutral indicator glowing yellow, a full turn to the right of the ignition key and, within two or three revolutions, the big twin burbles into life.

Are there any criticisms of the big Guzzi? This is difficult to say . . . one could pick on the fact that the fuel taps have no indication as to whether they're on or off and you might even comment adversely about the rocking pedal type gear lever, which takes a while to become accustomed to, but these are such niggling points that they cannot possibly distract from the host of points in favour of this giant of a machine.

Reputation

There's no doubt that those other transverse twins, the German BMW's, have built for themselves a tremendous reputation for smoothness and quiet running. The Guzzi V7 is at least their equal.

They are both aiming for a similar market —the hardened, but still conservative, big bike brigade and the long-distance tourer. Conservative though they may be, these riders still love machinery that is unassumingly different.

BMWs have provided this unique machinery for some considerable while, but now in the States there's the Guzzi and, unlike the BMW, the Guzzi draws the crowds of admirers. And this is its appeal.

It's one of those machines that doesn't need clip-ons or glass-fibre custom gear or four-pipe silencers. Straight off the assembly line, the Guzzi is without a doubt impressive enough!

How long will it be before we are able to buy the Guzzi V7 in Britain? As yet, no plans have been made to market it here—but many enthusiasts will be hoping!

REPORT AND PHOTOS BY JIM DAVIS

THE MOTO GUZZI V7 twin, was originally built to be a police motorcycle. When it was designed, due allowance was made for police accessories such as radio, footboards, windshield, sirens, and saddle panniers. As you know, a police machine must be able to accelerate rapidly, travel at extremely low speeds, sit and idle for long periods of time, and last but not least, give many thousands of miles of service before it wears out. Try to imagine what would happen if someone took a motorcycle like this, and made a civilian grand touring machine out of it. This is exactly what Moto Guzzi has done with their new V7 twin. The following test contains views and options after having the V7 for a little over two weeks.

Just about everything on this motorcycle is unusual. Best of all is the twin cylinder 90 degree V-type engine. It is an overhead valve four stroke, with a bore of 80mm and a stroke of 70mm. Total displacement is 703.717cc, with a compression ratio of 9 to 1. This big twin cranks out 50 horsepower at 6,300 rpm. Moto Guzzi, realizing that this bike would be used for police work, fitted hard chrome bores in light alloy cylinders. Many sets of pistons and rings should wear out before the cylinder bores will have to be changed. The cylinder heads are light alloy, and have a hemispherical combustion chamber. Lubrication is by the wet sump system just like in your family automobile. Because of this, no oil tank is needed.

As you can gather from looking at the photographs, the V7 has no chain, and secondary drive is by a double-joint driveshaft. Power is transferred from the engine to the transmission by a double plate, dry, automotive type clutch. The constant mesh gearbox has four speeds, and incorporates a cush drive that helps to add many more miles of use to the rear end assembly.

The frame is of the duplex cradle variety.

GUZZI V-7

AND NOW, A 700cc "MUSCLE MACHINE" FROM ITALY — A NEW CONTENDER FOR THE "SHAFT-DRIVE DEPENDABLE" CROWN!

BIG GUZZI handles like a lighter bike; has good clearance, even with low pipes. Not an oil drip anywhere on test unit.

Each forward downtube drops from the steering head, passes on each side of the generator shroud, swings along next to the sump of the engine, and then goes up behind the gigantic battery box and changes its loop to the other side of the bike under the seat. It looks like a lightweight but extremely sturdily-designed frame.

Suspension in the front is by a large set of telescopic forks, having an up and down travel of six inches. These massive units are just about impossible to bottom, but yet if they were allowed to extend rapidly the full distance, they hit the end of the stroke with a clank. Chances are that our test machine was a little low on oil, as another owner we spoke to said that he did not have this problem. The combination of the extremely soft dual seat and the large rear shocks was near ideal. The rear shocks are firm, but still just about perfect for a big bike like this one. The coil springs are adjustable, so if the ride is not to your liking, all you have to do is change the spring rate.

The top speed of our test machine was

LIGHTS at the base of speedo are: (l to r) high-beam, generator charge, neutral indicator and oil pressure light. Key has off, on and start positions.

DRIVESHAFT actually runs down middle of right swinging arm member, and is free to move up and down as the suspension moves. Two-plate dry clutch drives shaft.

ABOVE: Twin 29mm Dell'Orto carbs feed thru intakes that turn 90 degrees before entering amply-finned cylinders. Guzzi castings are first-rate work.

LEFT: One of the most beautiful engines we've seen; finned section is wet sump oil casing. Polished front casing covers generator pulleys, belt drive.

GUZZI V-7

an even 100 miles per hour. With a bike as heavy as this, and as fast as this, you have got to have a good set of brakes to bring it to a halt. The brakes on the Moto Guzzi V7 are just about the best internal expanding brakes we have ever had the pleasure of testing. The front unit is an 8½ inch dual leading shoe that has lining a full 1½ inches wide. When you put this brake on, you really know that you've grabbed a handfull of brake. Pressure needed on the hand lever is light, and with almost no effort the brake can be locked tight. The rear brake contains only one brake cam, but still, all it takes to lock the rear wheel at any speed is a light jab on the pedal. You've got to really try a set of brakes like these to see what we mean. Both the front and rear tires are mounted on beautifully finished Borrani alloy rims, that look sturdy enough to hold up a bike of twice the weight.

The electrical system is something else that you really have to experience to believe. The generator and electric starter are from an 1100 series Fiat automobile. Look at the photos with this test, and note the large box on each side of the bike that says V7, 700cc's. This box contains one of the largest 12 volt batteries that, we'll bet, you've ever seen on a motorcycle. No kick starter is incorporated in the V7, for the same reason that they don't build automobiles with hand cranks any more. Usually, all it took to start the bike was to set the choke half way on and flip the key in the headlight nacelle to the start position. Within a fraction of a second, the engine would leap into life, and then settle down to a magnificent throbbing idle. The only time we encountered any starting problems was one morning when we came out and found that we had inadvertently left the fuel tap on and the carburetors had overflowed and flooded the engine. All it took this time to start the engine was a fully open throttle and about two seconds cranking time of the starter. Not bad by anybody's standards!

Believe it or not, the electrical system is factory rated at 300 watts. This more than adequate for police usage and fantastic for the average motorcyclist's needs. A note to all you full-dress riders who like lots of lights on your motorcycles. The Guzzi V7 is the only way to fly.

The face of the speedometer looks like it was yanked right out of a car. It reads in miles per hour, and has an odometer. It has a group of lights (called idiot lights by many) at the base of the face. Seen from left to right, is the high beam indicator, the generator charging light, something that says idle, which is really the neutral light, and the oil pressure light. All these lights worked perfectly, but we're sad to say that the speedometer packed up and died at 812 miles. Prior to the time the poor thing died, we noticed that it had a tendency to give very optimistic readings in the true Italian tradition. No one would be silly enough to turn out a bad speedometer on a motorcycle as beautiful as this one so we're quite sure that we just managed to get the occasional "lemon" that crops up from time to time. On the same nacelle but below the speedometer is the key. It has three positions, which are off, on, and the spring-loaded start position.

For a bike weighing 539 pounds and having a wheel base of fifty-six inches, the V7 not only handles like a dream but goes like a real bomb. One of our lighter test riders was able to lift the front wheel a good foot in the air during our acceleration test every time he shifted from first to second. From second to third, if the engine was kept tightly wound, the rear tire could be forced to emit a satisfying chirp.

In the handling department, you'll have to look a long way before you find a big machine that reacts like this one. The center of gravity is very low, and we're sure that this has a lot to do with it. On a winding mountain road the engine produces so much torque that you very rarely have to use the gear shift lever. This leaves the rider free to concentrate on controlling the bike around the bends. This is one big bike that handles more like a scrambler than a grand touring machine.

We really dug the big V7. In fact, this was another one of these motorcycles that we really hated to give back to the distributor. If you're as enthusiastic as we are about beautiful grand touring bikes, you too will appreciate our feelings. Not only is this a beautiful, fast, and extremely agile motorcycle, but it makes us feel that its price is more than right. We suggest you take a look at one today. We would be willing to bet that you will agree with us wholeheartedly, when we say that the Guzzi V-7 is a great dollar value. □

Make	**Moto Guzzi**
Model	**V7**
Engine	**Overhead valve 90 degree V twin**
Displacement	**703.7**
Bore	**80mm**
Stroke	**70mm**
Compression ratio	**9 to 1**
Horsepower @ rpm	**50 @ 6000**
Maximum torque	**44 lb/ft at 3500 rpm**
Power to weight ratio	**10.8 lbs. per horsepower**
Carburetion	**Two 29mm Dell'Orto**
Clutch	**Twin plate dry**
Gearbox	**Four speed**
Suspension:	**Front: Telescopic; Rear: Swing Arm**
Brakes:	**Front: Double leading shoe; Rear: Single leading shoe**
Electrical system	**12V, 300W**
Starting System	**Electric, No kickstarter fitted**
Final Drive	**Shaft**
Gas Mileage	**50 mpg**
Fuel Tank Capacity	**4½ gallons**
0-60 mph	**7 seconds**
Top speed	**101 mph**
Price (FOB NY)	**$1439**

THAT CASING which looks like an electric starter is an electric starter. If you look under hood of Fiat 1100 car, you'll see the same 12-volt unit.

LEGEND OF THE V-8

MOTO GUZZI'S EXOTIC 1956-57 ROAD RACER IS STILL ONE OF THE WORLD'S MOST REMARKABLE RACING BIKES

By Richard C. Renstrom

ONE OF THE most fascinating stories in all of motorcycledom is the legend of the colorful and exotic Moto Guzzi road racing bikes. Famed for their speed from 1921 thru 1957, the cobby looking Italian machines won man thousands of races as well as fourteen World Championships; and the sport was certainly the poorer when the factory quit the Grand Prix scene at the end of the 1957 season.

The very summit of speed by this famous Latin concern came in the final two years of their racing campaign when they fielded the fabulous 500cc V-8 model. Not nearly so successful as the marque's single, twin, and four-cylinder models, the complex V-8 was, nevertheless, one of the world's great racing machines. It also was, and still would be, one of the fastest road-racing motorcycles ever built; and the romantic intrigue of the magnificent "eight" is still without peer in the annals of motorcycle sport.

The story of the unique V-8 began to unfold in 1954 when Ing. Giulio Carcano was watching the practicing for the Grand Prix des Nations at Monza. While lying in the meadow that day, Ing. Carcano timed the riders on the twisty sections of the circuit as well as the whole lap itself. Giulio was soon in a pensive mood, for he found that his shaft drive, in-line four-cylinder 500cc model was not getting thru the corners very well, yet its speed on the straights was no greater than the all-conquering Gilera "four". The 265-pound 350cc Guzzi singles were tops in their class, of course, but the fine handling 500cc thumper did not have sufficient flat-out speed to keep up with the "multis" on a course as fast as Monza.

All of this set the brilliant Carcano to thinking, for he realized full well that his "four" was a beast to ride and his big single could never be made to develop sufficient power to dust off the Gileras. Returning to his shop, Giulio sat down and thought about future development of a road-racing bike.

ABOVE: Guzzi used orthodox frame, but with oil contained in top frame tube. Exotic 'mill' punched out 75 HP at 14,000 rpm and was clocked at 178 mph in '57 Belgian GP, using 'dustbin' fairing.

RIGHT: Shown completely 'dressed', the Moto-Guzzi V-8 concealed much of its charm, hiding superb powerplant.

During the winter of 1954-55 Carcano turned his idea into a sketch, and then into actual mechanical drawings and specifications. The basic layout was a watercooled V-8 engine set crosswise in the frame so that a chain drive could be used, which was a radical departure from any accepted practice of the day. The principle involved was to reduce the bore and stroke to a minimum so that very high engine speeds could safely be used, which in turn provides a high power output. The new V-8 layout also provided great scope for future development, and by excellent designing it was possible to get a light and compact powerplant that could be mounted in an orthodox twin-loop cradle frame.

The first appearance of the new V-8 was at the Senigallia circuit in Italy, but teething trouble in the carburetion, ignition, and lubrication systems prevented the bike from actually entering the race. During 1955 the factory worked on the V-8 in an attempt to get more reliability, and by the spring of 1965 the model was given its baptism into international competition.

The power output in 1956 was reported to be 62 at 12,000 rpm, which was a little less than the 70 HP Gilera four. The bore and stroke of the V-8 was 44 X 41mm, which gave 62.31cc per cylinder and 498.48cc for the engine. The two banks of cylinders were at a 90° angle to each other, and a set of gears drove double overhead camshafts on each bank. The timing gears, which were located on the right side of the engine, also drove the water pump, with the radiator being mounted in front of the engine where the blast of air could easily reach it. The ignition system was by two distributors (driven off the left end of the inlet camshaft), eight coils, and a pair of 6-volt batteries. The firing order was 1, 8, 3, 6, 4, 5, 2 and 7; and tiny 10mm spark plugs were used due to the smallness of the hemispherical combustion chambers.

The crankshaft of the V-8 was a short and stout affair that was mounted by five split caged roller main bearings. The con-rods also used split caged roller big-end bearings, and the elecktron crankcase was cast in one piece. The cylinder liners were surrounded by a water jacket, as was the head castings. The valves, working in split guides, were set at an agle of 58° to each other, with ther sizes being 23mm inlet and 21mm exhaust. Two coil springs were used on each valve, and large tappets were fitted between the cams and valve stems. The domed three-ring pistons were deeply recessed for valve clearance, and they were shaped to provide a squish effect. Lubrication was by a dry sump system with a 9-pint oil reservoir being contained in the large diameter top frame tube.

Each of the tiny cylinders was fed by a 20mm Dellorto carburetor, with a pair of floats used to feed the eight carbs. A set of levers actuated all eight carb slides from one cable. The clutch, which was a multi-plate dry disc type, was mounted on the left side of the engine on the outside, while the primary drive was by a set of spur gears that ran in an oil bath. Gearboxes were produced with four, five, and six speeds, with the choice of a gearbox to suit the circuit. Carcano discovered, however, that the engine was so tractible at slow speeds that the six-speed box was not needed.

The frame used was quite orthodox, being a twin-loop cradle type with a swinging-arm rear suspension. One unique feature was the inclusion of the swinging-arm pivot in the rear section of the crankcase — a move to keep the wheelbase as short as possible. The front suspension was by the traditional Guzzi leading link fork with exposed shock units.

Moto Guzzi racers have always had huge brakes, and the exotic V-8 was no exception. The rear binder was cable-operated and contained in a large alloy full width hub. The front brake was a dual affair in a huge hub and with air scoops for cooling. The tire sizes were 3.00 x 19 inch front and 3.25 x 20 inch rear.

To finish off their speedy contendor, Ing. Carcano fitted a large eight-gallon alloy fuel tank and a rather bulky dolphin-type fairing. The dry weight of the V-8 at 320 pounds was actually quite light, but the bulk made the bike appear much heavier than that. The top speed on a 6.1-to-1 gear ratio was calculated to be about 162 mph, which was several knots below the highly developed Gilera four.

During the 1965 season the V-8 was raced by Bill Lomas (England) and Ken Kavanagh (Australia), but a great number of teething troubles prevented the "eight" from achieving any significant successes. In the Isle of Man the company mounted the team on their fine-handling singles; and in the Dutch GP, Lomas turned in the fastest practice lap on a

V-8 only to switch to a thumper for the race due to his bike losing its edge during the final practice session. Kavanagh retired while in fourth place, and he also retired in the Belgian GP as did Lomas while in third position. In the final event of the season at Monza, Keith Campbell of Australià replaced the injured Bill Lomas on the eight, but minor troubles again put the Guzzi out after the Aussie had lapped as fast as the leaders.

During the winter of 1956-57 the race shop toiled long and hard to make the mighty V-8 more reliable as well as raise more power. The first indication that the work had been successful came when the marque spent some time in the quest of several speed records during the fall and winter. The first to fall was the one-mile standing start, which Dickie Dale set up at 115.32 mph average speed on a course on the Ghedi Airport runway. This record is still standing in the F.I.M. record book. Bill Lomas also set up many speed marks, with his 151.4 mph average speed for the ten-kilometer distance from a standing start over the Terracina Straight, also remaining in the official F.I.M. record book.

During the middle 1950s, Grand Prix racing reached its peak in Europe, with many factories fielding works teams in the great classical races of the day. The year of 1957 is generally regarded as the summit of post-war motorcycle racing, and the competition between the leading marques was intense. During this period the Moto Guzzi was one of the leaders, and they naturally threw their heart and soul into making their V-8 a potential world champion. The results, however, were discouraging, since the V-8 experienced a string of disappointments that was certainly the opposite of the fabulously successful single-cylinder 350cc model.

The season started out well in the Italian Championship, which was won by Giuseppe Colnago against really top flight opposition. The famed Imola Gold Cup race was also won by the V-8, with Dickie Dale trouncing all of the world's great riders to prove that a winter's development had put more power into the high-revving "eight."

The first classic that year was the Isle of Man TT, where Dale experienced ignition problems but still managed to struggle home in fourth place at 94.89 mph —— well down from the 98.99 speed of Bob McIntyre on his howling Gilera four. Several fortnights later, in the Dutch GP, Keith Campbell pushed his Guzzi into second place, only to retire with a burnt-out clutch.

The next event on the international calendar was the famed Belgian Grand Prix — held over the Spa-Francorchamps circuit that is both beautiful and fast with its big sweeping corners and long straights that are set in lush green fields and deep woods. All through the night the fans poured into Spa until by morning the crowd was numbered at around 400,000, all eagerly awaiting the titanic clash between the many factories that fielded works teams in those days.

When the starter's flag dropped, the pack of Norton singles, Gilera and MV fours, BMW twins and Moto Guzzi singles and eights all shot from the grid and headed into the sweeping Eau Rouge corner. It was a beautiful sight, too, for the bikes had full dustbin fairings; and the sound of thumpers, twins, fours, and eights as they went thru the gears was a thing to behold.

The use of dustbin fairings gave a flat-out speed advantage over the dolphin type, and in a course so fast as Spa it was not surprising that the fours of Libero Liberati (Gilera) and John Surtees (MV) shot into an early lead. Keith Campbell on the screeching V-8 had other ideas, however, and on the fast Masta Straight he unleashed all the fury that his exotic powerplant churned out to take the lead from the fours.

After taking the lead Campbell settled down to proving who was really the maestro, for by then he had acquired

Continued on Page 79

TERRIBLY complex machine, the Moto-Guzzi V-8 featured eight carburetors, eight coils, two distributors, a pair of floats and water cooling. Crankshaft had five roller main bearings and the rods also had split roller big-end bearings.

CYCLE ROAD TEST

Moto Guzzi 750cc Ambassador

Moto Guzzi gives its big parade touring bike back to the designers to make it even bigger. Result: the "Ambassador" with both bulk and power.

PHOTOGRAPHY: JOHN SENZER

• Somebody would say it sooner or later, so it might as well be us. The new Moto Guzzi 750cc Ambassador is one helluva big, elephantine motorcycle. It's big in every department, from engine displacement to power output to physical size. It's big because that's what Moto Guzzi's designers wanted it to be. They built it around their concept of what a motorcycle should look like, how it should be used, how it should ride and perform. To our knowledge there is only one, maybe two, competitors in style to this bike—and both of them suffer distinct disadvantages. This is a bike that takes understanding, but when you know what it's all about, you might not settle for anything less. An elephant is a very serious animal.

Two of the goals behind all that built-in substance were obviously toughness and durability. Yet Guzzi's engineers knew that bulk alone wouldn't guarantee that their Ambassador would go 100,000 miles without major overhauls. What impressed us about this machine from the beginning was the beginning itself. We had arranged with Mike Berliner of Premier Motor Corp., which imports Moto Guzzi into this country, to pick up the bike at its New Jersey plant. Mike gave us that old show-biz routine that you used to read about in yesteryear's motorcycle rags. "Pick a crate, any crate," said Mike. "We'll unpack it right here and you can drive the bike home." We stared at this mountain of huge crates. "What color tank do you want?" he asked, "White, red, or black?" "Red," we said. "That column there," pointed Mike to the shortest column. "We shipped seventy of those last week." We pointed to a random crate and soon a fork truck had it out of the pile and onto the floor in front of his. One of the workmen pried the top and one side off and there stood an immaculate, gleaming, fully assembled Moto Guzzi Ambassador. All that was needed was to bolt-on the handlebars and fill the dry-charge battery. We watched in amazement when Heinz Kegler, Prunier's service manager, punched the starter button, blipped the throttle, and let the big twin settle down to a quiet, even idle.

The whole routine had a show-biz air about it but it was all very real. Here's what it means. Most motorcycle companies assemble their bikes on an assembly line (maybe or maybe not fire up the engines and run them) then disassemble the bike partially for packing into a minimum-size crate. That gives error a good chance at the action before you get your machine. During disassembly, parts may be lost or damaged. But more likely, the untrained junior mechanic who will unpack and reassemble

With its enclosed shaft drive, wet sump, and dry clutch, the Ambassador is very carlike in character—and it's equally durable.

that bike at your neighborhood dealer will get it half wrong, get the control cables out of adjustment, the right wire to the wrong socket, etc., and various nuts and bolts only finger tight. You'll have two or three months of adjusting, parts buying and debugging until the whole thing is working together like it should. At Moto Guzzi not only do they run the engine on a test bench for five hours before installing it in a frame, they also take the motorcycle for an exhaustive test run. Ours had 39 miles showing on the odometer. When they're satisfied they've got it right, they put it in a box, complete with oil in the sump, transmission and forks, and a new battery. They only drain the gas because law requires it. Our test machine was in perfect tune—*perfect*—the minute we fired it up. Five hundred miles later we had not made a single mechanical adjustment or correction of any kind. You see what we mean about serious?

Before leaving the Berliners' plant, we walked through the parts warehouse and saw a full stock of 750 parts, despite the fact that the first shipload of new models was still on its way to dealers. Ask some of your Far East manufacturers about their policies on that issue. Heinz told us to expect some torque reaction when blipping the throttle at idle, but that it shouldn't be noticeable at speed. This confirmed our experience with other transverse twins. He also said that at very low rpm there would be a slight transverse wobble as those heavy pistons oscillated in their cylinders. The cure, he said, is throttle. "Just get off briskly," he said, "You won't notice a thing." Getting off briskly isn't something you expect of an elephant. But we tried it and it worked... but more on that aspect later.

Guzzi's transverse engine, incidentally, is one of very few today in motorcycle production. BMW makes flat transverse twins, and so does Velocette for their little police motorscooter. In Russia, flathead flat twins are made for military use. The Guzzi, as far as we know, has the only "V" configuration, but like the others, it seems to have something military in its background. The Guzzis right now are used by Italy's army and police forces and are sold to the fuzz all around the world, including the U.S. We even heard a rumor that the Guzzi engine was first created for a small utility truck as part of a competitive bid to the NATO forces. We don't know how the bid fared and we couldn't get solid confirmation of the rumor. In any case, the Guzzi does have a military, even truck-like flavor. Massive, robust, heavy—these are the kinds of words that come to mind. It is also very functional, designed to run a long time with minimal maintenance. There's 300 watts of electric power to juice up those transmitters and sirens and flashing lights. There's a 6.1 gallon tank that doesn't need topping up every five blocks. There are plenty of handles and loops to grab and to strap things to; there's even provision for footboards if you prefer them to footpegs—as many police riders do.

The layout itself is unusual for a bike, but very automotive in character. Up front, a pair of heavy telescopic forks. The engine's two cylinders in a 90-degree "V" poke out to each side, just under the tank. The fore-and-aft crankshaft carries a large, automotive-type flywheel at the rear, which also acts as one bearing surface for the double-dry-plate clutch. A ring gear on the flywheel engages with the starter-motor coupling for electric start and no provision for kickstart is provided. The transmission, similar in many ways to that on the BMW, has three internal shafts, one of which has a spring-and-cam type wind-up shock damper. Power is not only geared selectively in the four-speed box, it also gets shifted to the rightside axis of the third or output shaft, where it couples, through a universal joint, to an enclosed driveshaft. The driveshaft turns inside the righthand fork of the rear swingarm and a pinion gear is attached to the end. This, in turn, meshes with a ring gear in the rear-drive housing, turning the rear wheel. All rear-drive components are supported in roller type bearings and run in a constant oil bath. Nothing here to wear out, nothing to adjust. Very similar to an automobile drive system except that it's totally enclosed and, because there is only one driven wheel, no differential is needed.

The 750 Ambassador is identical in most respects (except trim) to its V-7 700cc predecessor. New for the larger displacement and more powerful engine are cylinders, pistons, crankshaft, crankcase castings, and oil pump. Electric starter actuation has been changed from the spring-loaded automotive-type ignition switch to a remote pushbutton near the right handgrip. Moto-Guzzi has also fortunately dropped its rocker-pedal shift actuation for a more conventional single lever. Out of the box, the Ambassador now comes with right-foot braking and left-foot shifting (one up, three down pattern). If you prefer the British pattern, very effective crossover parts are available at a nominal cost. We like

If it dwarfs the ordinary motorcycle, it also dwarfs the miles. Start for Fairbanks on the Guzzi and you'll get there.

Why is this man racing a 600-lb motorcycle? Maybe in Italy they'll try anything. Anyway, he won the overland rally that favored lightweight

this interchangeability feature, for there will always be riders who want the option of change or who won't want to learn new tricks when they buy a new machine. One of the most important changes of all to the big Guzzi V-twin is its adoption of new, larger Dell'Orto carburetors with square slides and larger bores.

Dell'Orto concentrics have been around in various forms for over a year, but they are just now beginning to appear on production motorcycles. The one on the Guzzi is same model recently introduced on the Benelli 650 (still not available in the U.S.) The slide is definitely rectangular. Throttle actuation brings it up, pulling with it a metering needle that proportions fuel-air mix in midrange. A concentric float bowl on the bottom, similar to those on most Japanese carburetors, makes sure the fuel level in the float bowl is precisely right for the main jet no matter what the inclination of the carburetor. Result: no flooding or leaning out during hard cornering or on steep grades. The float bowl has also been designed to prevent ill effects from surging and hence flooding or fuel starvation during rapid acceleration or braking. But the best feature of all on this beautiful new carburetor is its accelerator pump. Because motorcycles don't have mechanical fuel pumps, they have always experienced carburetion difficulties when a rider wings open the throttle suddenly. Honda solved this problem on some models by adopting a vacuum-principle carburetor that opens according to the intake port's capacity to draw. Throttle cables actuate a butterfly valve, not the slide itself and hence the rider can never inadvertantly over-carburete. The problem is simply that a suddenly opened throttle presents less drawing pressure to the mainjet rather than more. Flow through the throat can't increase instantly. Dell' Orto's solution is equally clever. Through complex inner passageways, they use the sudden pressure created above the slide when it is pulled up to force atomized fuel into the carburetor throat. There is only one shot, but that's all it takes before normal throat-velocity catches up with the engine's demand. Snatch open the big Guzzi's throttle and the engine never gasps or falters. It leaps ahead like a horse given the spur. As well seasoned Heinz says, "Set off briskly."

Carburetion feeds through a short, curved manifold to the intake port. The

Guzzi's cylinderhead is all light alloy with shrunk-in iron valve seats. Valve size and valve timing (via rocker arms and pushrods to the camshaft) have been set slightly more toward the utility end of the useful tuning range than toward the maximum output end. Those 750 ccs deliver 55 or 60 SAE horsepower (depending on which manual you read) which is plenty of power but certainly not all that could be extracted from this engine. The compromise gives good, responsive midrange operation, with a broad, high torque curve. You don't have to row on the shift lever to ride this big machine. You get the feeling you could pull to the top of Pike's Peak at 68 rpm.

The cylinders are aluminum-alloy castings with hard-chromed bores. Perhaps this accounts for the stock roll bars fitted to the Guzzi. Most transverse twins use iron cylinders or alloy cylinders with iron liners. They won't dissipate heat as well as the Guzzi's alloy variety, but they do have more mechanical strength. A spill almost never knocks a cylinder off a BMW, but it might on the mighty Ambassador–except for the roll bars. A bonus from hard-chromed cylinders is that you never have to do a rebore. The pistons may eventually wear out, but the bores never will unless something sharp gets in there to peel up or score the chrome. Cylinders and pistons are *very* carefully matched at the Guzzi factory. As a military machine, it must be able to putt along for hours at three mph and cruise for hours at over 100-mph. That's a wide range of heat transmission to ask from an engine, and alloy cylinders and pistons with equal expansion rates and a high capacity for heat transfer are the answer. The fact that four rings are fitted to the piston indicates that the Guzzi's designers knew it would be operated much of the time at low rpm. Fortunately for the owner, it works well at both ends.

At the heart of the big Guzzi powerplant is its massive, forged crankshaft. Absolutely gigantic: you can't lift it with one hand. We doubt if any service will ever being needed for this piece of hardware. It turns in two huge main-bearings which are in housings bolted to each end of the main crankcase. In V-twin fashion, both connecting rods attach to the same crank pin. The rods use bolt-up end caps and split shell bearings. All these plain-bearings are kept alive through a high-volume oil pump that maintains about 60 psi at all engine speeds. Should a big-end bearing get too loose, you can replace it in automotive fashion from the bottom of the crank-

The only transverse V-twin in production–and this is how it looks after well over 600 miles.

Quick detach wheel keeps brake drum and parts completely isolated from oilbath gear drive.

If it's not a sprinter, the 750cc Guzzi doesn't hurt for speed. Run it up to 90 mph in third, with a gear to go.

case—without removing the engine from the frame. The camshaft rides above the crank in the cases and its plainbearings are also pressure-fed. All of these design characteristics are sound, conservative, and undoubtedly reliable. The overall unit, with its deep sump and generator on top is tall and heavy, but most of the weight (crank and flywheel) is down low where it exacts least penalty in handling. We noted with interest that a bash-plate has been fitted to protect the low-hanging sump.

That dc generator with its 300-watt output is the largest we know of in motorcycling. It was adapted from the Fiat 1100, and it is physically large and heavy. An alternator of equal capacity could be had at about half the size and weight, but the Guzzi's unit, made in so many thousands, is reliable and inexpensive. It is belt driven by pully from the front of the crank and a conventional voltage regulator keeps it from overcooking the battery. Back at the engine's rear, lower left side, another Marelli unit, the 0.7hp starter motor sits close to the battery. High amperage drawn during starting travels a short path and that is how it should be. Ignition, by battery and coil, is conventional except that the big Guzzi Ambassador uses an automotive type distributor. A V-twin naturally has irregular firing pulses, since the two pistons reach top-dead center asymetrically. The only fault we found with the ignition system was a high-voltage leak on the right hand side during rain. We noticed that the high-tension leads were very thin indeed, and maybe a pin-hole fault had occurred in one of them. In any case, if you reached for the fuel tap in the rain and overreached, you got at 20,000-volt jolt that you might not forget for days. This is not a systematic, but a local problem that could be cured easily enough by finding the leak and correcting it. The Guzzi's lights, horns and other electrics all worked splendidly with all that dc power on tap.

Engineers at Moto Guzzi took a surprising tack with regards to the new power offered by the "stretched" engine. Formerly 50 bhp at 6000 rpm with an 80mm x 70mm bore and stroke,

Continued on next page

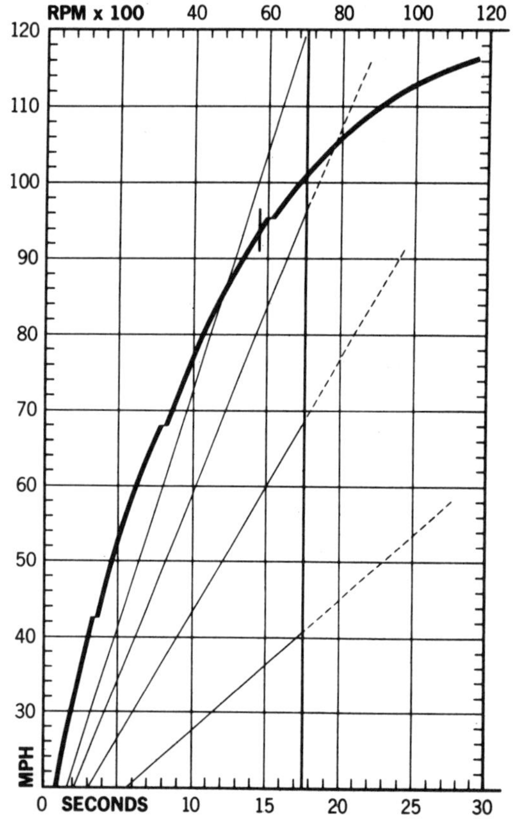

MOTO GUZZI 750 AMBASSADOR

Price, suggested retail	East Coast, POE $1539
Tire, front	4.00 in. x 18 in.
rear	4.00 in. x 18 in.
Brakes, front	8.66 in. x 1.66 in.
rear	8.66 in. x 1.66 in.
Brake swept area	90.1 sq. in.
Specific brake loading	8.43 lb/sq. in.
Engine type	OHV 90° V-Twin
Bore and stroke	3.27 in. x 2.77 in., 83mm x 70mm
Piston displacement	46.21 cu. in., 757.5cc
Compression ratio	9:1
Carburetion	(2) 29mm, Dell'Orto Concentric
Air filtration	Paper element
Ignition	Battery and coil
Bhp @ rpm	60 @ 6500
Mph/1000 rpm, top gear	17.1
Fuel capacity	6.07 gal.
Oil capacity	Wet sump 6 pt.
Lighting	Generator, 300 watts
Battery	12v, 32ah
Gear ratios, overall	(1) 13.415 (2) 8.018 (3) 5.738 (4) 4.511
Wheelbase	57.8 in.
Seat height	31.1
Ground clearance	5.9 in.
Curb weight	586 lb.
Test weight	761 lb.
Instruments	Speedometer, indicator lights
0-60 mph	7.8 seconds
Standing start 1/4 mile	14.36 seconds—93.16 mph
Top speed	116.1 mph

MOTOGUZZI

the new engine delivers 55-60 bhp at 6500 with an 83mm x 70mm bore and stroke—more peak horsepower and a wider power band. The Moto Guzzi designers decided to spread it out with altered gearbox and rear-drive ratios. Although the drive units are virtually identical, gear pairs have been changed. Now, low gear is lower (13.415 vs. 12.292) and high gear is higher (4.511 vs. 4.794). In street use this means better off-the-line acceleration and a higher top speed. Ordinarily the penalty would be in mid-range performance: the ratio-gaps between intermediate gears are wider, resulting in a certain sluggishness after each change-up as the engine has to work back onto its power range. It doesn't happen on the Ambassador simply because Guzzi's engineers tuned more for flexibility than peak power.

Big bike manufacturers have got themselves into a kind of craze for top-speed records lately, and we really don't know why. The BSA and Triumph 750s, the Honda 750—all boast of top speeds above 120 mph and they are all big, heavy bikes. BSA boasts the one-hour AMA certified world record at 127.615 mph, set early this spring. A few weeks ago at the Monza racetrack, Moto Guzzi got on the bandwagon with the one-hour FIM certified world record at 130.261 mph. (The FIM permits fairings, gearing changes and certain other modifications in the setting of production bike records). We don't question the veracity of these records and we suppose it's all done for publicity. In any case, we wouldn't drive at 130 mph on the public roads even it were legal. You *can* cruise at 80 on good roads if you stay alert and we suppose it's nice to know there's more on tap for passing. All of the big bikes have such performance—and so does the Ambassador.

How do all the pieces work as an integrated package? That's what you pay your money for and that's what every owner has to live with. Well, it goes without saying that this huge, elephantine Guzzi goes splendidly on the open road. Its mass as well as good steering geometry work together to smooth any bump, crack, seam or rough spot in the pavement. Gentlemen, that is a *big* motorcycle. An 80-mph crosswind won't phase it. It doesn't even quiver in the blast from passing semis. Majestically, massively it devours the miles. Plenty of power for the highway and you can run it up above 90 in third gear for quick passing. The word "smooth" has taken on a whole new meaning in the day of the multi. Certainly the Guzzi's 90-degree V-twin powerplant is smoother than any vertical twin. It's smoother than an out-of-tune multi. It has the kind of smoothness that lets you stay astride for as many hours as you can stay awake. Ride all day and half the night—you won't get numb on the Guzzi. Or ride it all night, as one of our editors did (he only got numb from the cold). He reported an engine-frame resonance at about 70 mph and so rode most of the night in third. By 500 miles the resonance disappeared and we felt that the Ambassador had become what it was meant to be: a solid, stable, powerful, and tireless highway cruiser.

A sprinter it is not. No 600-lb motorcycle is going to cut 13-second quarter-miles in street trim. The Guzzi 750 can more than get out of its own way, but you don't yank this one around like a polo pony. It has a big fat front tire and massive front forks. It also has a set of large and effective brakes—something forgotten on most two-wheeled highway heavies. It will corner at median speeds as steadily and firmly as a tightly-strung street-racer one third its size. But, like most of the heavies, it begins to drag its undersides in really severe cornering and it doesn't like choppy surfaces. A pity, really, because the "elephant" showed no other reluctance to handle the most torturous roads. At low speeds, handling was predictably heavy, what with those large forks and front tire, but when its throttle is wound out, the Guzzi seems to shed 200 pounds. In a straight-line, maximum-acceleration run, another handling idiosyncrasy shows up. When you declutch and grab for the next taller gear, a noticeable torque reaction tweaks the steering head as that heavy flywheel drops revs. This may be followed by another reaction as the clutch matches flywheel speed to transmission speed. It's all very quick, a little headshake or wobble, and then it's gone. The first few times you get a bit unnerved and find yourself afraid to shift when heeled over in a turn. But soon you learn to rely on the massive steadiness of the great Guzzi and on the realization that that headshake creates no compromise in control. We say it's not a sprinter, but it *will* crack off a quarter-mile in 14.36 seconds. That's as fast as most of the "lightweight" British 650s.

The Moto Guzzi Ambassador isn't everybody's motorcycle, not by a long shot. It's a specialty machine built for a specialty buyer. It won't satisfy the man who wants the very latest in technological evolution. It won't satisfy the man who wants a play racer or street sprinter. It will satisfy the man who wants strength, stamina and durability in a motorcycle, the man who has a long way to go and doesn't want to spend half of his travel time making adjustments or repairs. Quite unexpectedly, most of us learned to like the big beast, and the guy who rode it the furthest liked it the most. To a man, we came to respect it. Elephantine it is, but this bike has staying power. It cost $1539 (no set-up charge, remember?) FOB New York. For the big-bike buyer, it's worth every penny. ◉

LEGEND OF MOTO-GUZZI V-8

Continued from Page 72

sufficient experience to master the "eight" in braking and cornering. The Guzzi was also developing more horsepower then, with something like 75 HP at 14,000 revs being rumored. This much urge gave Campbell an advantage on the fast straights, and he continued to pull away from the pack at record speeds. On and on Keith sped, smashing the lap record every lap until he clocked 4 minutes and 26 seconds (118.56 mph). And then, just after being timed at an unbelievable 178 mph on the famed Masta Straight, the Guzzi came to a sudden halt — the victim of a broken battery wire.

This race, even if it ended badly for the Guzzi, was an historic occasion, for it was destined to be the last *Grande Eprueve* that the fabulous "eight" was to ever compete in. In the following classic at Ulster a rain caused the team to mount the finer handling singles, and then Campbell crashed in practice at Monza so that no V-8 was on the starting grid for the race. And then, at the end of the season, came the surprise announcement by F. B. Mondial, Gilera, and Moto Guzzi that they were all drop-

THE late Keith Campbell of Australia on the screeching V-8. Keith was '57 350cc World Champion, but never scored classic win on the V-8. Guzzi retired from racing at end of 1957 GP season.

ping out of world championship racing. The reason given was that the expenses had become too great for the factories to bear; and this, along with the F.I.M.'s decision to drop the use of dustbin fairings, signified that an era was ended.

Today, all that remains from that proud era are a few models rusting away in a factory shed at Abbadia Lariana in the shores of beautiful Lake Como plus a "show" 350cc model that has been exhibited all over the world. Designed by one of the great mechanical geniuses of our time and produced by a marque with a great legacy of exotic and successful road racing machines, the Moto Guzzi V-8 was certainly the very essence of what the grand prix is all about. As the Latins would say it, *Esto Perpetua* — may it live forever in the minds and hearts of men who revere and respect the magnificent saga of the Grand Prix.

The Moto Guzzi big single

THE MOTO GUZZI horizontal single is little-known in Britain although it is far from rare here in Italy, being used by the army and the police as well as by numerous private owners.

It is like the Vincent in a way, an "eccentric" machine, obviously the product of original thought.

Let me tell you about mine. I was out one day looking at an ex-W.D. Matchless G3L which was for sale (the garden sheds of Italy are full of W.W.2 bikes) and was allowed to have a ride on an old, allegedly 1947, Guzzi GTV.

It seemed immensely long and looked like a vintage steamroller with its external flywheel, underslung rear-suspension unit, low front seat and high separate pillion, exposed valve gear, etc. But it went stomping up the hills in top gear on its ridiculous 5.5 : 1 compression ratio and made a lovely thumping sound. Despite the vile hand-brushed paint and the non-functioning dynamo, I bought it for £25 the following week. Then began the work of restoration—not that I personally had the time, facilities, or skill to do much except change cables, sprocket, and chain. Most of the work was done by the local Guzzi mechanic.

These old singles are so durable and were built with gradual modifications for so long, that spares were never a problem. They just have to be still in stock!

The mechanic thought that the bike must have been built in the late thirties and re-registered after the war—he had never seen a post-war model with girder forks—but I wrote to the Como factory and they, in a very courteous reply, assured me that the bike was built in 1947 as the log-book says. They also sent me a free handbook from which I quote b.h.p. figures etc.

The GTV eventually emerged from the workshop, still with its evil paint, but now fitted with a new piston, valves, clutch, tyres, battery, primary drive (gears), wiring and silencer (which is a lovely mini-fishtail). The front forks, though still a bit sloppy, were passable, and the magneto and dynamo were reconditioned.

Since then the GTV has carried two people and luggage on a two-month trip all round Italy. We followed a devious route to Syracuse and back—3,000 miles without any trouble to speak of. I changed the plug in Sicily, used three litres of oil, and had to lubricate the front wheel bearing after one 100-mile bash over very dusty roads. This sort of reliability in a 22-year-old bike is impressive but perhaps not amazing considering the sheer weight of metal and the low engine stresses—not to mention the £50 overhaul.

It was an excellent bike for that kind of tour. Acceleration is unexciting and top speed an estimated 70 m.p.h. (no speedo fitted) but what more can you ask of 18.9 b.h.p. at 4,300 r.p.m.? The braking is pretty poor, too. The rear brake is reasonable but the front one is pathetic. The beauty of the beast is its quiet, effortless, 55 m.p.h. all-day cruising speed, utter reliability, and easy starting. The engine has considerable low-speed torque, pulls a 4.32 to 1 top gear and is beautifully smooth.

There is a vague sense of vibration somewhere down below but until the 4,300 revs limit is reached, when bad vibration sets in and the power falls right off, all the rider can really feel are torque pulses which are related to load not to engine revs. This smoothness is probably due to the horizontal layout and the stiff narrow crankcase permitted by the external flywheel.

Why has nobody else ever adopted the Vincent-upside-down rear suspension? Combined with the horizontal engine layout it results in a bike with an extremely low centre of gravity, which is unaffected by masses of luggage. Surely it's an extremely logical layout for a serious tourer? Certainly the rear suspension is streets ahead of the clashy front forks which tend to tie themselves in knots when the bike is being braked into a bumpy bend.

The 12-plate, mainly metal, clutch is very light but tends to drag. The gears are mounted in the rear of the crankcase. Unhurried changes are smooth enough but gearbox stiffness increases noticeably as the engine warms up, which may indicate an excessively high oil temperature in the gearbox. (Possible? They say Minis have the same problem.)

I wonder what happened to the Guzzis that Bob Foster imported in the 1950s. (See Titch Allen's article in March 1968 issue.) They would be the later models with enclosed valve gears, telescopic forks, a bigger front brake and hydraulic dampers at the back. They must be beautiful touring bikes with their improved handling and braking and another five or six b.h.p.

The Guzzi single, out of production for some years now, is about to be revived. A new version, retaining little of the original design except the remarkably oversquare dimensions (88 × 82), is about to go into production. Judging from the one unhelpful photograph I've seen, its frame wanders about rather less aimlessly than most previous Guzzi frames. I would love to try one.

Modena, Italy — *John Hine*

* I have just re-read your melancholy prediction of the end of big bikes (July 1968). I trust you will agree that things are improving. Recently the world-wide big-bike scene has been enormously invigorated by the arrival of Nortons that don't vibrate and may one day have a d.o.h.c. engine, the Tribsa threes, the Kawasaki three, the Honda four, the superb Laverda 750 twin, the bored-out Guzzi V7 and the Suzuki 500. All right, maybe the two-fifties, thanks to a better balance of power and weight, will eventually prove faster in racing, and admittedly you can cross London on a Honda monkey-bike quicker than on a Bonneville, but there will always be some riders willing to spend extra for high-torque cruising and an authoritative exhaust note. Anyway the Americans like "big-inch machinery" and what they want we'll get. J. H.

How's that for a long wheelbase . . . the 500 c.c. *Moto Guzzi GTV*

MOTO-GUZZI 750-cc AMBASSADOR

Moto Guzzi's Answer To Getting There In A Big Way, Ponderously, But Oh So Comfortably.

IF ROMAN GODS rode motorcycles, the Moto Guzzi would surely be the choice of Bacchus. It oozes pleasure and fleshy affluence. It is a big, stylish and strong road machine that reeks of luxurious plenty; a genuine ultra-bike.

The new Moto Guzzi is a good deal faster than last year's offering, but retains the plushiness necessary to a heavyweight, high speed tourer. Its strength lies in its quality and bigness. At rest, it seems to exude from every line and casting the quiet confidence that it will prevail.

This latest V-7 has been hopped up considerably.

The bore is 3 mm larger, while intake and exhaust valves are larger too, measuring 1.57 and 1.37 in. diameter, respectively. Also, a new type of square-slide Dellorto carburetor is used; the choke sizes are the same, 29 mm. The carburetors use accelerator pumps, giving snap to throttle response, but you must be careful when starting, as the engine can now be flooded by inordinate throttle turning.

These changes have appreciably increased horsepower, and the bike moves very well for one of such weight. Also, the rev limit has been raised 500 rpm to 6500 which seems relatively

PIZZA

tight for a luxury touring Twin. But designwise, this is actually a reasonable rate, as the very oversquare (83-mm by 70-mm) bore and stroke dimensions allow a piston speed in the region of only 3000 ft./min. at top rpm. This moderate speed is another factor ensuring reliability plus extended piston and ring life.

Our test bike's brakes are good; the rear unit can be locked almost at will, but is not unduly sensitive, while the front binder could benefit from more authority. It's strong when cold but tends to fade a bit when worked hard. The calculated brake loading figure of 8.0 lb./sq. in. would be strong enough for a substantially lighter machine, but for a bike scaling in at 559 lb. it's nothing spectacular. This is because brake swept area requirements don't vary in direct proportion to vehicle weight. Theoretically, of two motorcycles with the same brake loading figures, the lighter one should generally stop quicker.

Curious is the torquing effect exerted by the longitudinal crankshaft. It tends to lean the bike to the right as the throttle is blipped. The sensation doesn't bother the machine when in motion, although slight movement is perceptible at speed. But at a stop it is quite noticeable and possibly unnerving to the unfamiliar. Accentuating this is the generous heft of the twin disc dry clutch. It is actually more of an automobile type of assembly and adds considerably to the flywheel effect. But owners of BMWs, Indian Fours, Sunbeam Twins and others are not strangers to this phenomenon, and it all but disappears with familiarity.

Kicking over an engine this large could be rather trying for lighter riders. Fortunately, the factory has given the machine a 12-V, 0.7-bhp electric starter which never fails to light it off promptly. Apparently this is a faithful unit as there is no kick lever. Have no fear of overtaxing the battery, however, as the generator is a belt-driven 25-ampere, 300-watt Minarelli charger. Incidentally, the battery is rated at 32 amp/hours and weighs more than 28 lb., well up to almost any contingency. If your Guzzi were fitted with a windshield, it would even be feasible to power a set of wipers for wet touring. Egad!

The crankshaft is a one-piece forged steel unit, the connecting rods sharing a common big end journal. The rear end of the crankshaft rides in a mammoth 2.124-in. plain bearing. The front main measures somewhat smaller, at 1.495 in. These are impressive dimensions, even for an engine of this displacement. It's obviously built to last. Interesting is the crankcase design. It is of a one-piece configuration, with the crankshaft accessible from the rear. Beside the benefits of facilitating teardowns and service, oil leakage has been minimized with the elimination of seams.

The alloy cylinders, rather than incorporating iron liners, have chromed bores. Along with cooling advantages, the chromed bores are extremely hard, giving very good wear characteristics. Cooling is aided because the heat has to travel through a fraction of an inch of chrome, rather than a thick, iron liner. Nor is there a boundary between a liner and cylinder for the heat to cross. The chrome bore also allows piston-to-wall clearances a good deal closer than iron cylinders. Moto Guzzi literature indicates that the pistons should be fitted with just .0021-.0025 clearance. Leaving the pistons such little room to rattle around greatly enhances engine life and keeps down noise.

Connecting rods are a conventional steel two-piece pattern, but the pistons are unusual for motorcycle components as they mount four rings. Two are for compression and two for oil control, with the bottom ring shortly above the lower edge of the skirt. Although four rings are seldom seen in modern motorcycles, the factory opted to maximize oil control and longevity and minimize piston rocking in the bores.

Acceleration for a machine this size is excellent. Our test machine, with fully suited test rider, weighed a grand total of 725 lb. The V 7 didn't mind this at all, going through the Lions Drag Strip clocks at 91.88 mph in 14.74 seconds! Power output is rated at 60 bhp and test results bear this out as a totally honest figure. Engine torque is abundant, and the flywheel effect is fierce! Second gear shifts brought almost a foot of daylight between front tire and pavement while third gear chirped the rear tire. Quarter-mile starting technique called for full rpm at the line and unceremoniously dumping the clutch. This consistently resulted in loud protests of the tire as it burned 30 feet or more out of the chute, but the big Moto Guzzi just did its thing, imperturbably, time after time. The twin disc dry clutch became a bit fragrant after a while, which is understandable, but it never slipped.

Throughout the test, CW riders missed not one shift. Going through the gears particularly at lower speeds, the transmission "clunks" quite definitely. But gear finding was never difficult, even in drag race haste. Shift lever action was reasonably smooth and lever travel not excessive.

While discussing the shift lever, we should mention that Moto Guzzi dealers offer conversion kits to allow right-side shifting and braking on the left, instead of the standard arrangement. The ratio spacing between first and second gears is not so noticeable when upshifting, but the second-to-first downshift should be made at speeds less than 5 mph because the rear wheel tends to skid under the sudden compression braking. We suspect that the real villain is all that flywheel effect which prevents the engine from gaining revs quickly.

As discovered earlier, the machine's appearance belies its capabilities, and handling is no exception. It's no lightweight and doesn't lend itself to being stuffed into a corner, but it is very nimble for its size. At low speeds it is clumsy, but once the bike is underway clumsiness fades away and it feels quite agile. Upward of 30 mph, the V 7 seemingly loses about 200 lb., and at freeway speeds it is even better. Vibration at low rpm is prominent, similar to sitting on a large water buffalo with a grumbling stomach, but smooths out as the machine gets underway.

It is very easy to swing the bike up on its centerstand, free from hernial repercussion. In fact, it's less trouble than many 250s we've come across. There's nothing gimmicky about the centerstand, only that proper effort has gone into the design of a part all too often overlooked on other bikes.

The handlebar/seat/footpeg layout is comfortable enough for all-day touring and, together with the Moto Guzzi's high speed handling virtues, make this one of the most desirable touring mounts we've ever encountered. Two-up riding is also very accommodating, particularly in the 60-70 mph range where the V 7 is extremely smooth with much more muscle still on tap.

With the introduction of such a large, durable and powerful motorcycle it's only natural that many police departments throughout the country would be greatly interested. Indeed, many of them are. In fact, it was the interest of the LAPD that gave the impetus for the 750 Moto Guzzi. When the first V 7s debuted in 1967 the Traffic Division tested several for freeway duty. The final consensus was that the bikes needed more steam to cope with the weight of all that extra equipment. Thus the hot new 750, of which the LAPD is currently evaluating 10 units.

Understandably, for CW staffers, this has been a pleasant test to conduct. The machine proved always willing to take its rider almost anywhere at a moment's notice, and do it right smartly. It is a powerful and elegant, if somewhat bulbously styled, high-speed cruiser made for wide open spaces.

MOTO-GUZZI 750-cc AMBASSADOR

SPECIFICATIONS

List price	$1554 p.o.e. West Coast
Suspension, front	telescopic fork
Suspension, rear	swinging arm
Tire, front	4.00-18
Tire, rear	4.00-18
Brake, front, diameter x width, in.	8.66 x 1.66
Brake, rear, diameter x width, in.	8.66 x 1.66
Total brake swept area, sq. in.	90.1
Brake loading, lb./sq. in.	8.05
Engine, type	four-stroke ohv V-twin
Bore x stroke, in., mm	3.267 x 2.775, 83 x 70
Piston displacement, cu. in., cc	757.5
Compression ratio	9:1
Carburetion	(2) 29-mm Dellorto VHB
Ignition	battery and coil
Claimed bhp @ rpm	60 @ 6500
Oil system	dry sump
Oil capacity, pt.	6.15
Fuel capacity, U.S. gal.	6.07
Recommended fuel	premium
Starting system	electric
Lighting system	battery, 12 V
Air filtration	paper element
Clutch	twin disc, dry
Primary drive	gear
Final drive	shaft
Gear ratios, overall:1	
5th	none
4th	4.51
3rd	5.73
2nd	8.01
1st	13.41
Wheelbase, in.	57.5
Seat height, in.	32.0
Seat width, in.	12.8
Handlebar width, in.	32.6
Footpeg height, in.	11.5
Ground clearance, in.	5.8 (at center stand)
Curb weight (w/half-tank fuel), lb.	559
Weight bias, front/rear, percent	50/50
Test weight (fuel and rider), lb.	725

TEST CONDITIONS

Air temperature, degrees F	78
Humidity, percent	58
Barometric pressure, in. Hg.	30.05
Altitude above mean sea level, ft.	350
Wind velocity, mph	6-8 mph
Strip alignment, relative wind:	

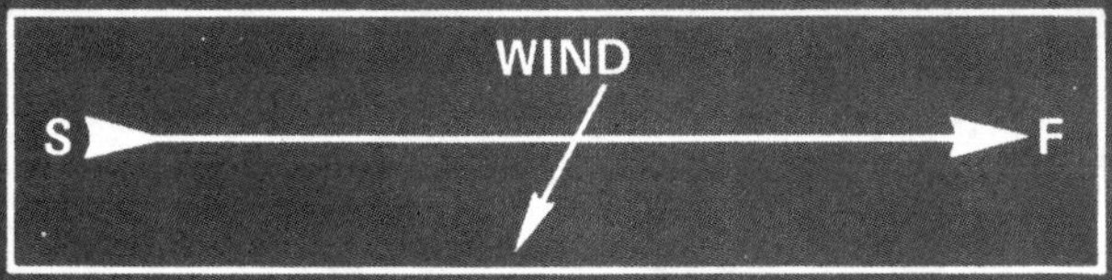

PERFORMANCE

Top speed (actual @ 7000 rpm), mph	115
Computed top speed in gears(@7000rpm),mph:	
5th	none
4th	115
3rd	91
2nd	65
1st	38
Mph/1000 rpm, top gear	16.5
Engine revolutions/mile, top gear	3645
Piston speed (@ 7000 rpm), ft./min.	3237
Fuel consumption, mpg	46
Speedometer error:	
50 mph indicated, actually	47.26
60 mph indicated, actually	57.14
70 mph indicated, actually	66.81
Braking distance:	
from 30 mph, ft.	43
from 60 mph, ft.	148
Acceleration, zero to:	
30 mph, sec.	2.2
40 mph, sec.	3.4
50 mph, sec.	4.6
60 mph, sec.	5.7
70 mph, sec.	8.0
80 mph, sec.	11.0
90 mph, sec.	14.2
100 mph, sec.	21.0
Standing one-eighth mile, sec.	9.37
terminal speed, mph	75.46
Standing one-quarter mile, sec.	14.74
terminal speed, mph	91.88

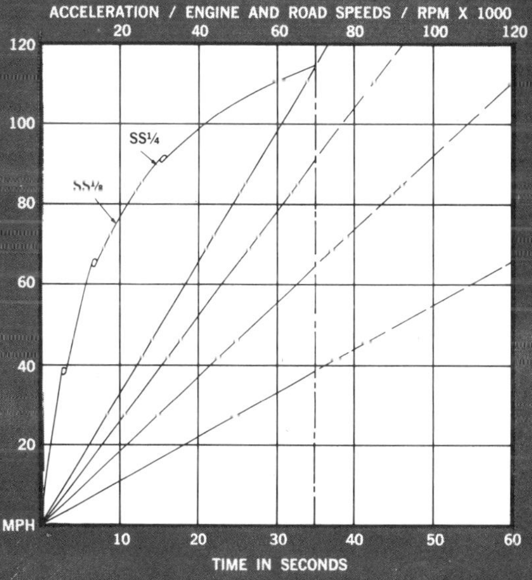

750 MOTO-GUZZI

NOW A 750cc MODEL, THE MOTO-GUZZI V750 AMBASSADOR CATERS TO THE TOURING RIDER AS WELL AS THE POLICE, BEING OFFERED IN VERSIONS FOR BOTH

MOTORCYCLIST TEST REPORT

ABOUT A YEAR AGO we tested a Moto Guzzi V7. In looking over our old test we saw that we liked the bike. It had good brakes and seemed to be a real possibility on the grand touring scene. It was as reliable as a motorcycle could get, delivered good gasoline mileage and was comfortable. Really, this is all that you can expect from a touring motorcycle, so when the Guzzi people claimed that they had added 50 cubic centimeters to the engine, improved gasoline mileage and made it go even faster, we were interested enough to grab

ABOVE: Dual test — 750cc Moto-Guzzi Ambassador is also available as police model (at left). Both are compared in accompanying test report article.

RIGHT: Another view of fully equipped police model. A number are in use in police departments in such key cities as Los Angeles, New York and Atlanta.

MOTO GUZZI AMBASSADOR TEST

one for a test. The following is what we found out about the new 750.

The V750 is fast for a bike weighing in at a solid 600 lbs. We clocked 111 miles an hour at one of the local tracks when our machine had only 480 miles on the odometer. This means that the claimed factory top speed of 115, once the bike is broken in, is more than a faint possibility. One thing that just about shook us out of our saddles was the fact that this big, heavy bike will either spin its tire in second, or lift its front wheel on a shift to second. This, of course, will depend on the skill of the rider, but can you imagine enough torque from an engine like this one to "pop a wheelie" at the snap of a throttle?

The Los Angeles Police Department and several other PD groups around the country have purchased a batch of these machines in conjunction with their other motorcycles, and as we wondered why they liked this unusual machine enough to do this, we had the West Coast distributor, ZDS Motor Corp., drop off a police bike along with the "civilian" model. We're glad they did, as we enjoyed the police bike as much, or more, than the other machine.

At low speeds the V750 is a heavy handler, as can be expected because of its weight. Once you get the momentum up, all that instantly changes. The Guzzi will never be a road-racer, but at the same time is as light as any other big bike on the road. Much of this comes from the excellent suspension, and beefy frame. Front suspension is heavy-duty tubular telescopic that seems more than capable of handling even a bigger bike. The rear set-up is standard swinging arm controlled by oil-damped shocks. It works well, but what is unusual about the shocks is that they can be adjusted to the amount of weight you're carrying by grabbing the plastic ring at the bottom of the shock, and turning. You don't need any sort of special tool, and the entire adjustment job only takes moments. The double downtube, cradle-type frame is composed of long flowing bends, large gussets and tabs to hold all the goodies in place. Look at the size of the engine and you'll see why the frame had to be beefy.

The bike is absolutely too much, but it's the engine that immediately tells you that the Guzzi is not an ordinary motorcycle. As you can tell from the rocker covers, it's a four-stroke overhead valve engine, but look at the way the engine is set into the frame. Transverse mounted opposed twins aren't so rare, but transverse V-twins are. One Japanese manufacturer built some of them a few years ago, but had difficulty selling them. During the Second World War, a major U. S. manufacturer built quite a few for military use, but for some reason didn't continue them for too long after the war.

Piston displacement of this monster is 757.486 cubic centimeters. Bore and stroke are 83mm by 70mm, which means that the engine has to turn some rpm to get its power. Would you believe that the engine is rated at 60 horses at 6500 rpm? From the bore and stroke dimensions, you can see that the engine is a "buzzer", but when you ride it, it just ain't so! The bike will lug away from a stoplight with the engine sounding like an oilwell pump running at slow speed. The specs say buzz the engine, but riding it says torque, torque and more torque. Compression ratio is 9-to-1, but here again, the engine is so docile feeling that you'd guess that the compression was somewhere around 7-to-1. As you can see, most of the engine is made from light alloy. The cylinders are aluminum with hard chrome linings. The cylinder heads are aluminum, internally hemispherical in shape and are equipped with cast iron inserted valve seats. Pistons are also aluminum as is the entire crankcase.

Connecting rod bearings are babbit-type instead of roller or ball, but as the engine is far from being stressed, it doesn't matter. Stop and figure that your automobile has the same type of bearings, and it will run for at least 100,000 miles if you treat it with a little care. Here again, the lubrication system is similar to many sportscars. All the internal components are pressure-lubricated at 54 to 60 pounds, and the system is wet sump, which means that the oil is stored in the crankcase instead of a separate oil tank. Carburetion is by two 29mm Dell'Orto "square carburetors" with the left one being equipped with an accelerator pump.

Those of you who know sportscars will recognize many of the electrical components of the V750 as coming from a well-known imported car. We won't tell you which one, but the next time you see one of the smaller Fiats, compare the Guzzi starter to that of the Fiat. This ought to give you a good clue.

Before we go any further, let's climb aboard the civilian Guzzi and take a short ride. We'll get up front, and you ride pillion. Right off the bat, you notice that large soft seat. Kind of reminds you of a Vincent seat, doesn't it? Pop the rear folding pegs down, hold on and let's go.

On the old V7s the key on the nacelle used to control ignition and starting. On the newer model, you flick the key to "on", and punch the button on the right handlebar. If the engine is in neutral, it will roar to life in a turn or two. We found from experience that the neutral light on the speedo face can't be trusted – at least it couldn't on our test machine. Before you get underway you've got to slightly rock the machine to make sure the engine's out of gear. Once the engine is running, stick your toe under the long shift lever, pull in the clutch and lift up. The bike will go into gear with a satisfying clunk. The reason for the clunk is that the clutch is similar to that of an automobile. It's not a single-plate assembly as in another well-known motorcycle, but instead has two dry-type twin-driven plates. Engagement is positive, not harsh, but at the same time produces a clunking sound. The plates do not run in oil, but run dry, exactly as in an automobile. We let the clutch lever out and the bike pulls smoothly away. For a big bike, acceleration is far above average. In high gear you can actually feel the punch when you're running at 100 and crack the throttle. Around town the bike feels light until you get the speeds down, and then you realize you're sitting on a big heavy bike again. To shift, stick your toe under the lever and lift up. For the other three gears, push down on the lever. You'll find neutral in between first and second.

The engine produces so much torque, that four speeds in the cogbox are more than enough. Our test bike wasn't even slightly broken in, but yet it would putt along at 20 miles an hour in high gear, and then accelerate to 111 miles an hour in high gear.

Check the brakes in this set of photos. The front is a gigantic dual leading shoe unit that looks large enough to stop a grand prix bike. The rear is only single leading shoe, but yet could not be persuaded to fade unless badly mistreated during several hard panic stops. We were impressed with the braking of the new "super" Guzzi. It didn't have disc brakes or any other sort of exotic stoppers, but yet the tried and true internal expanding grabbers with all their surface area did the trick every time.

Drive to the rear wheel is not by chain, but by shaft drive. The specifications state: By constant speed double joint cardan shaft. Bevel gears ratio (transmission-rear wheel): 4,375. What this means to the average rider is that he won't have to adjust the chain, the chain won't throw oil mist all over his girl-friend's clothing, and the whole thing is as foolproof as a motorcycle drive can get.

CLASSIC styling of Guzzi V750 Ambassador. With 600-lb. weight and long wheelbase, model is well suited for highway touring.

Our performance specs just about matched those shown in the sheet we were given with the bike. They were so close, in fact, that they only differed by one or two miles an hour. What's amazing is that the police machine that carried almost 200 pounds of extra equipment and goodies came within one or two miles an hour of these specs also. Thru the gears, our speeds were: 1st: 32 mph, 2nd: 62, 3rd: 83, 4th: 111 mph.

Maybe we shouldn't do it, but would you like to know something about the police version of the Moto Guzzi? At least when a man in blue with the blinking red lights pulls you over, you'll know how he caught you. First, we'll tell you that for a big scoot, the thing goes like whoopee. In first, it'll burn rubber till you let off the go valve, and in second, it will wheelie just like the pretty red and chrome model. One of the members of the motorCYCLIST staff had the pleasure of finding out that his stolen motorcycle was recovered after a high speed chase by a motor officer riding one of the new Moto Guzzis. What more can you ask?

The machine we had as a police test was unusual in many respects. To begin with, it has the loudest electronic siren we've ever had the displeasure of hearing. We hadn't even thought of setting it off, but some little grubby kid saw the bike in front of the office, flicked a few switches, pushed a

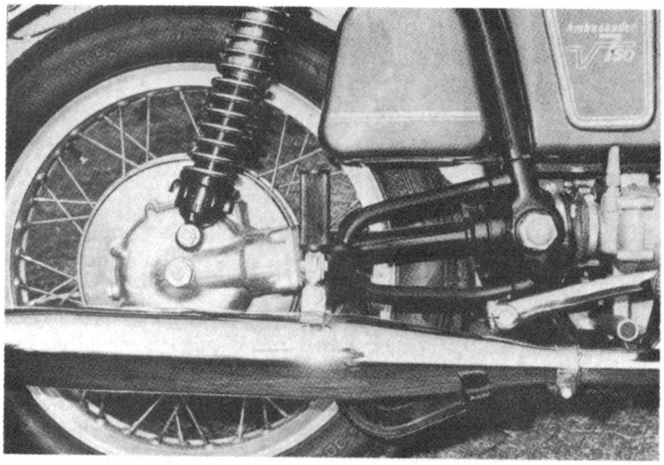

REAR shock units are 3-way adjustable. Drive shaft is enclosed in swinging arm unit.

ELECTRIC starter, used in some Italian-made cars, is shown. Beneath it is foot shift pedal.

few buttons and got a little more than he had tried for. When the racket had subsided, the little one had wet his pants, and we had collected just about every police car in the area. If it means anything to you, the siren and loudspeaker combination is produced by Steward Electric of Los Angeles.

We hate to admit this, but most of the members of the staff actually liked the police machine more than they did the "civilian" version. Both bikes are almost the same as far as specifications go, but the fixtures bolted on are quite a bit different. Those of you who have, or remember the early V7s know that the shift lever was on the right side. It's now on the left on both kinds of bikes. The police version has footboards, and because of this has been fitted with a rocker shift lever and a modified brake pedal. The combination of solo seat, footboards and long, swept back "sit up and beg" bars makes for one of the most comfortable riding motorcycles we've sat on for quite some time. We felt that the handlebars on the stock model were a little too short, but other than this, we just couldn't find any fault with the bike. Seeing the police boards, pedals, levers and handlebars have to be kept in stock, we see no reason that you couldn't order a bike set up like this on special order.

We could go on and on about the new Moto Guzzi Ambassador II, but we're running out of space. If you're considering a motorcycle for long distance touring, and you desire comfort and prestige, check into this bike. We're sure you'll like it.

SPECIFICATIONS:
MOTO GUZZI AMBASSADOR V750

Displacement	757.486cc
Engine	OHV, transverse mounted twin
Bore and Stroke	83 x 70mm
Compression Ratio	9-to-1
Horsepower at rpm	60 @ 6500
Top Speed	111 mph
Carburetion	29mm "square" Dell'Orto
Clutch	Automotive type, double driven disc
Gear Ratios	1st: 13.415, 2nd: 8.018, 3rd: 5.738, 4th: 4.511
Brakes	Front: Double leading shoe Rear: Single leading shoe
Frame	Tubular Double Cradle
Weight	600 pounds
Length	88.6 inches
Ignition	12 volt battery
Lights	12 volt battery, 300 Watt generator
Starting	Electric
Price	$1554 FOB Los Angeles

ENGINE	Excellent	Good	Fair	Poor	Unsatisfactory
Starting	•				
Throttle Response	•				
Vibration	•				
Noise	•				
TRANSMISSION					
Gear Spacing	•				
Clutch Smoothness		•			
Shifting speed		•			
CONTROLS					
Handlebar position	•				
Ease of operation	•				
Location	•				
BRAKES					
Lever pressure	•				
Pedal pressure	•				
Fade resistance	•				
Directional stability	•				
Stopping distance	•				
SUSPENSION					
Front	•				
Rear	•				
Ride control	•				
Dampening	•				
APPEARANCE					
Paint	•				
Construction and welds	•				
Chrome and trim	•				
ELECTRICAL					
Wiring	•				
Headlight	•				
Taillight	•				
Horn	•				
GENERAL					
Spark plug accessibility	•				
Instrumentation	•				
Side stand	•				
Center stand			•		
Seat comfort	•				
Muffling	•				
Tool storage space	•				

COMMENTS: A quality motorcycle. Excellent brakes, good handling and more than enough power for touring on the expressways in the United States.

A DIFFERENT BREED OF CAT

The chance of winning is slim, but everybody will know he was in town.

This year at Daytona, there is going to be a lot of scrapping among the factory teams to win the big one — the 200 miler that comes on Sunday. Everybody is pulling the stops out — all the big guns will be there riding the latest machinery that has been carefully assembled in secret by the world's leading factories.

One unfortunate thing about Daytona is that everybody is so preoccupied with all of the factory-sponsored entries that all too often the poor privateer who comes to Daytona on his own hook, and spends a great sum of money, is often overlooked. If it were not for the privateers motorcycling would be a dull thing indeed, with only a handful of riders providing the show.

One of the most interesting and intriguing machines we have run across this year originally started out as a 1969 Moto Guzzi Ambassador. The two people responsible for the creation of this unusual machine are Bob Blair of ZDS Motors (who, coincidentally, happens to be the distributor for Moto Guzzi) and George Kirker (who also just coincidentally happens to work for Bob). They pooled their talents and time and have come up with a motorcycle that is not only unusual, but beautifully executed. It would be difficult, if not impossible, to estimate the fantastic number of hours that went into the construction of this machine, although we are sure it is considerable.

While George and Bob combined efforts, it will be George who will be doing the riding, and he's well qualified for the job, too, because for the last couple of years, George has been the production racing champion here on the West Coast in the Big Bore department. He even managed to pull off a couple of wins in the G.P. class on a production type motorcycle. This is no small feat when considering the talent he's running against.

The Guzzi itself has been pared down to 331 lbs. dry weight, which is certainly competitive. The day these pictures were taken we went out to Orange County Raceway, and George suggested the editor take a couple of laps, just to see how the bike felt. The way the bike was geared, the top end was only 120 miles per hour, but the bike would lift the front end in second and third gear when shifting with the throttle wide open. It pulls like the proverbial freight train and our estimate would be speeds over 150 in road racing trim, geared properly for Daytona.

There has been a lot of talk, both pro and con, about the effects of torque with an in-line mounted crankshaft. When sitting on the Guzzi and blipping the throttle, you definitely feel the torque

trying to move the bike from one side to the other. This force reaction is also felt when exiting a corner. Depending on whether it is a right or left hand corner, application of the throttle tends to pick up or push the machine down slightly. While this is no big thing, the rider does learn to compensate for this very quickly, although for the first couple of corners it is a bit disconcerting. George feels this is an asset rather than a liability, and he can use this to actually help him rather than hinder him. We certainly hope so.

Instead of a full fairing, Bob and George have fitted a half or partial fairing for our photographic exercise. With a full fairing it would be difficult to see all of the beautiful handwork involved in this motorcycle.

The bike is anything but peaky. It accelerates like a dragster, which proves once and for all that the Guzzi is definitely applicable for other things than touring, or riding to work.

The original drive system is retained, since it has worked out very nicely. Upon close examination you will discover the driveshaft running from the gear box to the rear wheel passes through the swinging arm. It's a neat little trick and does away with a separate driveshaft housing.

A DIFFERENT BREED OF CAT

The engine itself is quite interesting. There were no lower end modifications whatsoever. It was felt that the engine was strong enough as is, and nothing that Blair or Kirker could do would improve reliability. The valve train had been reworked with a Norris cam, springs and rocker buttons. Along with the modified induction system, 10:1 Mondial pistons were installed. The heads were taken off, flow tested and reworked. West Coast tuning wizard, C. R. Axtell, did the job, and at Daytona time, the two square slide Delorto carburetors will be replaced with injectors. Pete Jackson is making up a special pump and injection system just for this motorcycle. It should be quite interesting to see how this works out compared to a standard carburetion system.

The standard exhaust system has been modified and Ducati megaphones are installed. By the way, while on the subject of the megaphones, we might add that the sound is unlike any other four-stroke twin we have ever heard. The exhaust note is so distinctive that the spectators will have no trouble telling George from the other participants.

When it came to the ignition system, they stuck with the stock advance curve, with the timing set back from 10° to 8°. At present the ignition system is being replaced with a magneto for Daytona. For shorter races, the standard battery ignition is fine, but for something as long as Daytona, a magneto is much more desirable.

While there were no modifications to the lower end, it was balanced by Edelbrock and the smooth running Guzzi is even smoother running than before.

While the stock frame was used, much grinding and lightening took place to reduce the weight as much as possible. The rear section of the standard frame was removed and a brace added between the shock mounts. While the stock frame could be retained, stock forks could not. They were not only heavy, they were not up to the job. A pair of Rickman forks and a Lockheed disc brake are used on the front end, and they really do the job. The forks work quite well, and the brakes are really great. They pull the rider down from 120 (at least as of this writing) very quickly, and it really doesn't take more than a couple of fingers on the front brake lever to accomplish this.

Standard shocks are presently being used, although we would venture to say on the rough banking at Daytona George may want to go to something a little bit better. Time will tell.

In conjunction with the Lockheed disc and the Rickman forks, a 19" wheel with a 300 × 19" Dunlop is used on the front and a 400 × 18 Dunlop is used in the rear.

If the fuel tank in the photographs looks familiar to you, it should. It came from a Triumph that ran at Daytona, but had to be shortened and notched to fit the new frame. Blair and Kirker did this themselves.

Since the engine is a wet sump design, there was no oil tank needed. Our personal feeling is that we would rather see a dry sump engine with an oil cooler because at Daytona that engine is going to really get hot.

A seat pirated from a Triumph racer is used, shortened four inches and notched for the rear shocks. It looks as though the saddle was originally made for the Guzzi. The boys at ZDS did all of the painting and upholstery too, we might add.

Before we forget, we would like to mention that Steve Zabbaro did all of the welding of which there was a considerable amount. It's amazing, in a project of this type, how many pieces of metal have to be stuck together with the aid of a torch.

George did the paint work and Von Dutch did the striping and lettering. All of the workmanship and detail are first class. There are no fuzzy little corners usually found on even the most expensive racing machines. It looks as sharp and is as well prepared as the most expensive factory racing equipment.

The only thing about the bike that we didn't like was the transmission. It was quite "clunky" and not as smooth as we would have liked it. George mentioned that a good portion of this problem is attributed to the heavy flywheel, and a new light aluminum one is being manufactured at this very moment. Smooth shifts can be made, providing the rider lets the engine revs drop before engaging the next cog, either going up or coming down. This, of course, might be fine for a touring bike, but for a racer, it is impossible to live with this situation. The light aluminum wheel should definitely help things.

Over the past several years we have run across a number of rather unusual competition machines, but we think it would be quite safe to say this would probably be one of the most, if not the most, unusual bikes to come along in quite a spell.

It's doubtful that George will win, but we're quite confident that he will be among the finishers, and one of the rules of racing is that you have to be around at the finish to win. Finishing is important — far more important than blinding speed and acceleration, good handling, fantastic brakes or anything else you want to mention. Reliability and stout-heartedness are two things that must be inherently bred into any racing motorcycle, especially one that is going to have to pound around Daytona for 200 miles. We are going to be there this year, and while we don't think George will win, we wouldn't be a bit surprised to find him in the top ten finishers, which, for a first attempt, is almost phenomenal.

This Guzzi cum racer proves that you need not have a pile of money to produce a potential winner. Ingenuity, enthusiasm and painstaking workmanship will go a long way toward getting that laurel wreath and fat check.

"Moto Guzzi's well-deserved reputation for good handling . . . this machine is all that one would expect"

One of the few 125s that seem to be a "real" motorcycle

WHEN they told me I had to collect a 125 Moto Guzzi to test, I scratched my head. "The 125 Guzzi ? Isn't that the one with the desmodromic valves ?" "No, that's the Ducati." I pondered a little. After a few more inspired, but futile, guesses I finally confessed that I did not know what it looked like. Neither, it seemed, did anyone else and I arrived at Barrett of Redhill with a rare treat in store. It is not often that I don't even know what the bike I am going to test looks like!

Redhill, on the borders of rural Surrey, is not the sort of place where one would imagine a motorcycle business would thrive. Horses, maybe, but motorcycles ? Maybe Redhill isn't *really* a motorcycle area, but Don Barrett, "gaffer" of Barrett of Redhill, has taken it by the scruff of the neck and made it one. The premises in Station Road are typical of those of hundreds of motorcycle dealers throughout the country—small and busy-looking. The difference is in the machinery in stock. In the window the day we arrived a Moto Guzzi V7 Special and a Kawasaki Mach III vied for attention, while a smaller Guzzi, a Ducati and a trials Montesa basked in the reflected glory. Just round the corner near the station, in what seemed to be a giant showcase, were a dozen or so 125 Guzzis, Yamahas, Ducatis and Bridgestones, all new, with that very rare bird, the Moto Parilla, nestling in a corner with its mate. As well as a couple of road-going versions of this interesting machine there were a few sporting versions, scramblers and racers. Barretts are concessionaires for Moto Guzzi and Parilla and have more than a passing interest in Kawasaki and the Montesa trials iron. We like to see a dealer backing the sport of motorcycling and we were pleased when he cheerfully admitted that business was good. He had a full workshop and was having a job to cope with the demands on his servicing facilities. How different from a dealer who constantly cries fings ain't wot they used to be.

When I don't know what a machine looks like I tend to build a picture in my mind of what it *ought* to look like. I expected to find a sleek Italian-looking model with clean engine, smooth lines and flat bars. I was wrong about the bars and also about the seat which I thought (feared ?) would be one of those painful planks Italian manufacturers seem to have a habit of fitting to their small machines. The rest of the machine was pretty well as expected. The engine is one of the smoothest-looking we have had in a while, with a cylinder barrel that would not disgrace a 250. The line of the machine is as neat and balanced, as only the Italians seem able to contrive, with every item part of the machine rather than seemingly tacked on as an afterthought. The whole machine looked and felt just right except for one thing—the upswept handlebars. We will be charitable and assume that the bars are really intended for the USA market. Almost every manufacturer seems to think that Americans want high-rising handlebars no matter what the machine. We think that they are wrong. The Moto Guzzi is a stylish, sleek roadster that handles like a racer. It not only would look better with a set of flat handlebars—it *needs* them. Fitting upswept bars to a machine like this is like fitting clip-ons to a Harley Davidson. It has nothing to do with comfort, for there could be no criticism on that score; it is purely aesthetics. Some small bikes can get away with upswept bars, some they even improve. On the Moto Guzzi they jar.

All this, of course, did not detract from the pleasure of riding, and in most respects we felt this was an excellent machine. It was one of the few 125s we have tried which seemed to be a real *motorcycle*. Other machines in this class, rightly perhaps, have made no pretence at being anything more than functional A to B vehicles. They aim for the commuter market and hit it smack in the middle. The Moto Guzzi, on the other hand, stands there proudly looking exactly what it is, a proper motorcycle. It is not the approach that will win over many "non-motorcyclist" riders, for they are going to dismiss it as too sporty for their needs. Which is a pity, for it showed itself to be as good for town work as any other machine in its class, with the added bonus of being fun to ride.

The engine used in the Guzzi has a 52mm bore and 58mm stroke, giving a swept volume of just over 123 c.c. It has a compression ratio of 9.8 to 1 and produces a healthy 12 b.h.p. The engine is of extremely robust construction to deal with this respectable power and has a massive crankshaft assembly. If there is a production 125 with a bigger assembly than this I have never seen it. It would fit into a 175 without looking undersized. The big-end has roller bearings and the small-end a plain bush. The crankshaft is supported in the crankcase by a ball bearing at each end with the 28w alternator mounted inside the right-hand end cover on the end of the shaft. The helical-gear drive transmission is in an alloy case on the left-hand side and consists of four gears, one for the crankshaft, one for the gearbox incorporating the clutch assembly, another for the camshaft and the last one (which is not part of the primary transmission train) for the oil pump. A four-ring piston is used with two rings and a scraper above the gudgeon-pin and another scraper at the bottom of the piston. The carburettor is mounted unconventionally to the left hand side of the engine and inducts into the side of the head *via* a fairly long tract. This enables the valve gear to be mounted across the engine and the pushrod tubes to be placed at the rear of the barrel. It is not clear whether this is to tidy the arrangement of the valve gear up or to increase induction efficiency. As a result, the exhaust pipe comes out pretty well from the right-hand side of the engine. It certainly makes for a very tidy top half. The valve gear is enclosed in a neat alloy cover and retained with Allen screws. The valves run in an oil bath and a breather runs from the top of the engine to the rear of the machine. The entire engine unit was 100 per cent oil tight throughout the test.

The clutch, a five-disc unit in an oil bath, is, as stated, in the left-hand transmission case but the final-drive sprocket is the other side of the engine and drives to the rear sprocket on the right-hand side of the machine. The engine is of full unit-construction and the gearbox shares its oil supply with the engine. There are 3½ pints of oil in the sump to perform this task and the only real criticism we could make is that the dipstick, mounted at the front of the crankcase, needed a spanner to remove it. As a result, the temptation not to bother to check is considerable. Apart from that the whole engine is so well constructed that it is difficult to imagine it suffering unnatural wear.

The engine unit also forms part of the frame structure. Twin down tubes bolt to the front of the crankcase. There are no frame members underneath the engine and the frame is then bolted to the rear of the engine at the top of the gearbox, this being the rear swinging-arm sub-frame. Twin tubes also pass from the top of the swinging-arm units to the rear of the sump.

The telescopic front forks were softer than we expected, but the rear dampers reminded us that the Italians like their

All-alloy 52 × 58mm engine unit, with pushrod-operated valves, is set at 25 degrees and is claimed to produce 12 b.h.p.

Seventeen-inch wheels, 50*in wheelbase and* 2¾ *gallon tank contribute to "full-size" compression*

suspension firm! Although the rear units had exposed springs they had sensibly placed dust covers inside the springs. If we *must* have exposed rear springs it does seem to be a good idea to prevent the ingress of road dirt into the very sensitive suspension units.

The petrol tank size is a very respectable (for a 125) 2¾ gallons, with a reserve of just under ½ gallon operated by a second tap. The filler cap is an elaborate construction with a hinged lever holding the cap in place, racing style. Unfortunately it wore out the chrome underneath! The dualseat is large and was comfortable, as were the fat rubber handlebar grips. Perhaps one of the more attractive features of this little machine are its alloy castings. Other people are satisfied to produce mere castings, Italians make them works of art. The handlebar clamp, rocking gear-change lever and rear-brake pedal are all good examples of the craft of the alloy worker.

When we called for our test machine we were first offered an identical used model from stock for a few days as the new bike was not quite ready. What an excellent idea this turned out to be, for very often one rides just one example of a particular machine and wonders if an annoying fault or good point is typical. Our two machines gave us the chance to find out.

The first Guzzi Sport had done over 5,000 miles and was well run-in, the second had under 400 miles on the odometer. Both were identical in specification and colour and both felt just about the same to ride until it came to performance. The old model was well run-in and was a flier and easily recorded, and exceeded, the claimed 70 m.p.h. top speed. The new machine clearly needed longer to loosen up and would only approach the claimed top speed in extremely favourable conditions. We were quite satisfied that it *will* reach the claimed speed and were glad of the chance to try another model. (We wonder how often performance figures are freaks, one way or the other, on road-test machines ?)

Starting the Moto-Guzzi meant first of all forgetting it was a 125. That 9.8-to-1 compression ratio has given venom (as in snake!) to the kickstarter, and timidity of approach earned one a sharp rap on the shin. One soon learned that a firm swing would bring the engine instantly to life with a deep throaty crackle. The silencer did its job well enough but still allowed a sound of substance to leave it. I grew very fond of the exhaust note on this machine. It could best be described as rather flat, somewhere between that of a Tiger Cub and the old 250 Aermacchi, with the emphasis moving away from the Cub the faster one went. It was a lovely noise, not at all what one would expect from a 125. It appealed so much that I suspect it might have clouded my judgment a little, and perhaps it was a shade *too* loud for non-believers.

To add to the exhaust we had a healthy clatter from the alloy motor. It is to be expected that such a large area of alloy would transmit a degree of valve noise, and it did. The sound was one of tappets with three times the gap they should have but as both machines transmitted the same sounds it must be taken as typical.

The motor was ready for action as soon as it is started. An air control lever is fitted to the top of the carburettor but was not used in the mild weather we had during the test. Wet, maybe, but mild. Assisted by a first-rate gearbox the engine unit showed a lively interest in getting a move on. The gear-change really was a peach, so light and quiet and with such a short travel that it was difficult to accept that the change had been accomplished at times. The penalty for this is that, with such a light touch needed, one sometimes journeyed between first and second a few times in a vain search for neutral. The rocking gear-change was useful for keeping one's shoes clean but, as we always seem to do with such systems, the rear half of the arrangement was soon ignored.

For years Moto-Guzzi have had a well-deserved reputation for the handling of their mounts, and this machine embodies all that one would expect. The hard rear suspension, counterbalanced by a comfortable dualseat, functions well enough to avoid discomfort but maintained a firm grip on the activities of the rear end and the machine could be cornered with the greatest of enthusiasm. I was *really* trying before I succeeded in grounding

Continued on next page

MOTO-GUZZI 125

one of the tapered rubber footrest rubbers. Only the handlebars inhibited the cornering capabilities of this machine for, instead of the leisurely flick which one would have needed with smaller bars, a positive effort was needed. One surprising aspect of the Guzzi was the steering lock. Not the type that makes the machine thief-proof—it didn't have one of those. The machine could be turned in a circle that would do justice to a trials bike. About the only merit of the large bars was that they encouraged such activities.

The Guzzi was quite happy to settle to 55/60 m.p.h. indefinitely, and this was achieved with the motor sounding surprisingly unhurried. Engine rattles vanished and all that could be heard was the thud-thud-thud of the exhaust. The addition of a pillion passenger made very little difference to performance, although the passenger was soon complaining of the position of the pillion rests. Vibration, that curse of the riding classes, was undoubtedly present, although it could only be felt if one gripped the petrol tank with the knees. The shape of the tank was such that this was not a natural position and it was relatively easy to sit with the knees just off the tank.

The speedometer is one of the least attractive features of the machine. It is a headlamp-mounted unit of the type one usually sees on mopeds and seemed quite out of place on this sporting machine. It was also given to flattery.

The 25/25w headlight was powerful enough for moderate use provided the engine was spinning well but suffered from a badly-spread beam. It seems a pity that the 7-amp-hour battery has not been put to better use, for it supplied just the parking light and h/t coil. The biggest drawback to the lights was the way they fell with the revs, and with the engine ticking over the light disappeared almost completely. The horn was acceptable—but could have been better.

The brakes were excellent, a model of what brakes on a lightweight should be. Assisted by the very good Pirelli tyres, the machine could be stopped very rapidly, even in the wet. The rear brake suffered from its rear adjusting nut coming undone, due presumably to vibration. (Could be nasty . . .) A reasonable toolkit shares space under the seat with the battery. This space also acts as a "filter" for the air intake. No filter as such is fitted, and the arrangement used would seem only to serve to prevent the ingress of the large foreign bodies.

Finished in grey, with black frame and red and black tank, the 125 c.c. Moto-Guzzi Sports endeared itself to us while it was in our possession. It was quick, reliable and fun to ride. It also had a heart as large as its cylinder barrel. In spite of being driven with great enthusiasm for almost all its stay with us it recorded 125 m.p.g. overall. Selling at £175, the Guzzi suffers only from a lack of publicity. Which is a pity for it is a very pleasant little machine. B. P.

ROAD TEST

Continued from page 47

the front of the engine making the engine part of the cradle. Although some vibration was a result, it seemed no more than any other high-compression single of this size.

Ground clearance at 7.87 inches is adequate but not ideal, especially with the abbreviated bash-plate. It's a small slatted steel structure connected to the downtube, leaving the crankcase uncomfortably vulnerable from below.

If you like the appearance of functional machinery, the Guzzi rates high. Its grey-painted diamondesque muffler is starkly attractive, as is the tuned sound it emits. The workmanship and finish are very good, resisting the abrasive effects of gas, water and grime applied liberally and with great vigor during our tests. And nothing fell off. There is a multi-base paint wherever paint can go, leaving a minimum of chrome exposed to the corrosive environment of trialing.

A well-stocked tool kit is included and from all appearances Moto Guzzi has tried to make servicing easy. For instance, changing tires is facilitated through permanently fixed tommy bars on the wheel spindles.

The shapely Italian-styled gas tank wore a competition-type, quick-open gas cap. Capacity was 2.46 gallons with ⅓ gallon reserve. Engine feed could be monitored through clear neoprene fuel lines. Oil flowed nicely into the crankcase sump and from there via a gear pump, into the engine, clutch and gearbox components.

In the headlight nacelle are a miniscule speedometer/odometer, a red main light indicator, green parking light indicator, and the ignition aperture.

Though we didn't have to use the brakes much they seemed adequate, but not superb. The front brake actually came closer to our likes than the smaller rear drum. Another departure from the usual trials setup.

All in all, the Moto Guzzi ISD Trial is a delightful, eager performer that looks good; aesthetically *and* economically. But it should never be taken for granted or regarded as a drab, dependable workhorse. A thoroughbred is developed for the sport. It must be exercised, curried, coddled and broken to your riding habits. Look elsewhere for a dray horse; look here for excitement. If you want to cover your bets across the board, drive the 125 Trial off the beaten track. ◉

MOTO GUZZI I.S.D. TRIAL 125cc

Price, suggested retail	FOB East Coast $589.00
Tire, Front	2.50 in. x 19 in. knobby
Rear	3.00 in. x 19 in. knobby
Brake swept area	52.8 sq. in.
Brakes, front	7.5 in. x 1.38 in.
rear	5.75 in. x 1.13 in.
Engine type	Four-stroke single
Bore and stroke	2.043 in. x 2.283 in.; 52mm x 58mm
Displacement	7.515 cu. in.; 123.11cc
Compression ratio	11.4 : 1
Carburetion	22mm Dell 'Orto
Ignition	Alternator and coil
BHP @ RPM	13.5 @ 6800
Fuel Capacity	2.38 gallons
Oil Capacity	½ gallon
Gear ratios, overall	(1) 34.2, (2) 24.0, (3) 15.8, (4) 10.3
Wheelbase	49.21 in.
Seat height	31 in.
Ground clearance	7.87 in.
Curb weight	222 lbs.

GUZZI'S 750 AMBASSADOR

The Ultimate Tourer?

MOTO GUZZI have never been great ones for tradition, and the general run of motorcycling's peculiar lore and order. While the rest of the world goes barmy in its attempt to keep up with the Joneses, by producing machines identical in all save details such as tank shape and colour, 'Guzzi stick to *their* traditions. Just try and pinpoint a design blanket policy such as our own factories follow, i.e., the vertical twin mania, or big single blindness, both still worshipped by some people as the ultimate in motorcycling. Like the Gadarene swine, manufacturers and public alike herded together, their senses numbed by their own cries of righteousness, and stampeded straight over the precipice. Luckily, a few held back, (whether or not it was due to quick thinking or slothfulness, is debatable) and from them we struggle back to sanity, attempting to claim the throne that was once ours, but is now too far out of reach.

'Guzzi have been pioneers of a multitude of different engine designs, frame styles, suspension types, and styling, few of them ever resembling their contemporaries' products. Not all have been successful, especially internationally, for whereas the home country knew its product, the rest of the world shied from the apparently unknown. Sound, world-wide marketing was all the biggest Italian manufacturer lacked.

They broke with their own traditions of uncompromising, unconventional engineering and design and turned conservative, to offer the world a machine utterly steeped in the well-worn paths of proven automotive engineering slap, bang in the heart of the Harley country, with whom it is successfully duelling. Every aspect of the 'bike has been exploited to the full by other manufacturers at one time or another since time immemorial. No new ground has been broken. Everything has been done before. Ideas have been refined and sophisticated, but the Ambassador is, to all intents and purposes, an old-fashioned motorcycle. Here Moto Guzzi genius shines through. *It has never been done before!* No-one has picked out the ultimately most reliable motorcycle performance features, and built a machine from them. 'Guzzi have done what so many others have been doing for years, and soundly and roundly beaten them at their own game. The result is magnificent.

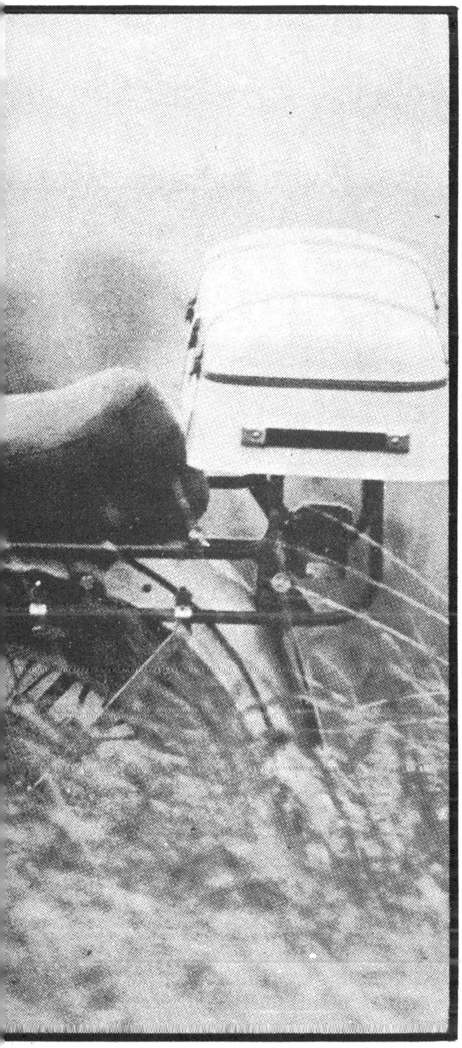

What will be the effect of machines such as these on our own manufacturers upon our entry into theCommon Market? £900 may be the British price just now; in Italy and the rest of the EFTA countries it's about £600, plus delivery charges. That's about the same as the BSA/Triumph threes and the Commando.

Not all of this tradition is confined to the narrow sphere of motorcycling of course, as should be obvious by the construction of the Ambassador; many of the engine and transmission characteristics have been lifted from car practice. Every individual component or feature has at some time or another been incorporated into a 'bike, but the whole is reminiscent of a car's engine. Indeed, the complete electrical system is identical to those used on many of the smaller Fiat cars. This was a quite deliberate design feature.

This particular machine, on test, belongs to Roger Woods, an ex-Vincent owner, and that is significant, and, in some ways, surprising, considering the view of most Vincent owners of anything that is not from Stevenage. For many years Roger and I ran Vincents together. I was luckier than he, and enjoyed a more reliable, if slower, 'Shadow. It was all to do with bad rebores and Mk II cams, but more than enough to ensure Roger's dissatisfaction with his machine's reliability, although not its performance.

Every marque presents a personality entirely its own; most have something in common with each other. In my experience, the Ambassador is alone, offering its owner something no machine has in the past, or does at the present time. (With possibly one exception). Sheer, unabashed luxury, even at slight cost to other, lesser performance, details. Although it is a fast motorcycle, ripping acceleration and a high top speed are obviously regarded by its makers as of less importance than comfort, silence, reliability, and high cruising speed ability. Very few motorcycles nowadays fall into the category of tourer; the 'Guzzi certainly does.

Silent, relaxed riding

REGARDLESS of the machine's or engine's speed, mechanical silence was exceptionally good. Roger has fitted a Tri-Point windscreen, and this exaggerated all aspects of noise from the machine by echoing it straight back at me; but even so the only discernable noises were from the tappets, and transmission gears on the overrun. With an air-cooled engine nothing can be done about tappet noises, and with no windscreen fitted, I doubt that the criticism would have arisen from either source. Until the experience of riding a motorcycle with no chains at all, anywhere, in its construction, has been enjoyed, the delights of truly silent, relaxed riding cannot be fully appreciated.

Starting was a little noisy; rather like a car's which is not surprising. With the button pressed, the electric motor whined, clashed with the flywheel teeth, and excitedly chattered the grumbling big twin into life, which it did in a slightly longer period than the Honda Four, or a car, but just as reliably. Cold starting needed full choke. Without it, the firing strokes remained out in the cold. For a mile, full choke was necessary, then within the next five, the handlebar lever had to be gradually eased to zero until the engine's temperature had risen enough to take over. That it took such a time was probably due to the two alloy cylinders' estranged relationship, and their open air life, plus the fact that the large alloy engine unit must absorb considerable heat. Left standing for only 15 minutes, starting would again necessitate use of the choke. Wisely, Moto Guzzi have precluded the use of a kick-starter. On an engine of this type, the extra complexity of a bevel geared crank, unless of the BMW type, would be unnecessarily expensive, complicated, untidy, and heavy.

Weight was quite obviously regarded by the designers as all part of the sacrifice to luxury, and I must admit to agreeing with them. A couple of enthusiastic friends of mine with a personal interest in the Laverda's success, or otherwise, take an understandably defensive attitude regarding the Ambassador. If you want a shaft drive you buy the 'Guzzi, or if a chain, then the Laverda; it is as simple as that. The two are similar machines. Their reaction is that of Bugatti to Bentley's grand creations. "He makes the fastest lorries in the world", or something along those lines. I leave it to the two owners' clubs concerned to continue the fight, but regarding the 'Guzzi, as with other tourers, I just do not think it matters, being one of those strange people who are prepared to accept weight as a necessary characteristic of modern, powerful, fully equipped, motorcycles.

Acceleration was deceptive, principally because of the uncanny silence of the ferocious movement forwards, once its initial inertia had been overcome, and the power was rolling smoothly in just above 5000 rpm. A top speed of 109 mph was recorded. At 60 mph the speedometer was four mph fast. Assuming the error continues further up the scale, for speedometer errors rarely if ever correct themselves, the Ambassador's ultimate speed in my hands was approximately 106/109 mph.

A speed wobble was caused by a Dunlop K70 on the front wheel, fitted, after a spot of tyre swopping, when the standard Pirelli on the rear began to throw its tread. A 70 mph lightness, oversensitivity, disappeared when I accelerated, and remained hidden until the machine slowed near top speed, then returned in strength, fourfold. The shimmy turned to a barrelling, which travelled through the machine and started the 550 lb monster swaying and lurching rhythmically from gutter to gutter. At top speed, accelerating out of trouble is out of the question so I slammed on both brakes hard, and accelerated instantly. The movement stopped; then I did too. I can only assume that the tread of the K70, being non-directional, started the business, and then the weight of the machine distorted the deep carcass, which upset things even further. Roger hopes to solve the problem with a TT100. Perhaps Dunlop would comment? Incidentally, the trouble never showed up under acceleration, or bend swinging.

GUZZI'S 750 AMBASSADOR

In the lower two gears, travelling along at 1500 rpm and even less was possible, and 2000 rpm in the higher ones, thanks to the flexible engine power and positive transmission; 2500 rpm in top gear (4.51:1) represents 34 mph, and at the other extreme, 1500 rpm in bottom gear (11.62) represents 9 mph. Usable acceleration came out of the lower rev-ranges mentioned, but the advantages of sticking to the higher ones ensured that a far more powerful movement resulted as the throttle was used. More than enough to deal with town traffic. At a little above 5000 rpm, by maybe 100 revs, no more, a slight power increase helped the machine along more eagerly, so for maximum performance each gear change was made at 6000 rpm, ensuring that no gearchange dropped the revs below that. Providing the rev-counter and gear lever were used with this in mind, a distinctly sporting performance could be encouraged from the lazy, pushrod engine, that otherwise suggested strongly that the best way to travel was with the minimum of gear-changing, and an easy throttle hand. It was too. No adrenalin gland twitcher, this 'bike. Stick it in top, leave it there unless forced to change, caress the throttle, persuade it but never snap it to the stop. Unless you happened to be in that sort of mood of course, and then, as I said, the ride really does give you something to think about.

Maximum revs were marked at 6500, but from the feel of the engine I suggest that the 'Guzzi would have taken another 1000 before protesting, not that it is relevant to the final performance. Gear speeds, at 6500 rpm were; 1st, (11.62:1) 41 mph; 2, (7.59) 65 mph; 3, (5.73) 86 mph; 4, (4.51) 106 mph, but at 6200 rpm. The machine is low geared, perhaps too low in top; It really needs five ratios to make the most of the performance; top gear could then be raised beyond its present level without destroying the ultimate fourth speed gear. Otherwise gear spacing is excellent. Bottom being low, one on its own, and the others grouped closer together above it for easy changing, and top allowing maximum speed at maximum revs. The clutch had to be used; without it, changes were rough, and clunky. Cruising speeds were anything up to 100 mph (5900 rpm), but the Ambassador was at its best and most comfortable around 85/90. At that speed, behind the windscreen, I relaxed completely, out of the wind, in silence, and watched the world drift past.

Beyond other standards

ONLY one other vehicle I have driven exceeded the comfort of the 'Guzzi, and that was a 3CV Citroen. It, the Guzzi, was head and shoulders beyond the standards of any other machine, and beyond that of any cars as well, save the wee Citroen. Leaning out a little, I watched, fascinated, as the front wheel rippled and bounded between the forks, yet passing on none of it to the machine. The seat was huge; a great slab of well-shaped, deep foam rubber, it was wide as well, and so comfortable I forgot its existence completely. The five-gallon fuel tank blended perfectly with the seat and my knees. It held enough for a non-stop run of 250 miles at 70 to 90 mph – my speed – 55 mpg. Driven hard, consumption would presumably fall to the level of other 750s – 30/40 mpg.

For such comfort, small sacrifices must be made; in this case to the machine's optimum roadholding ability. Not handling, mind, that was perfect, but roadholding was. . not. . quite. Grounding happened too easily, so after finding out the limit, I left it alone (would you push things on someone else's £900 motorcycle?). What surprised me was the leech-like way the monster stuck to a line on a fast bend. One or two I know up in Fen country go on forever. Round and round the 'Guzzi would go, faster and faster, while I sat cradled in its big, strong lap. The headlamp and control panel a long way forward, peering, immobile, ahead of me. I did not sit on the 'Guzzi, but in it, and so, cradled in luxury, sped. The fault came up a few times on these long, fast bends. An unexpected camber change, an undulating bump, or a side-road camber extension, set the 'bike yawing; a movement, while not dangerous, was merely unexpected, vaguely disturbing. If the strong damping required to iron out the phenomena was applied however, then the Ambassador's comfort would be destroyed along with its air of graciousness. Until a new suspension system is designed, we cannot have it both ways. Me? I'm for the 'Guzzi way. Certainly, I was not slowed at all by the movement.

Line chopping was exceptionally good, equal to a far lighter machine's abilities. Apart from other design qualities, much of the credit must go to the Ambassador's low centre of gravity. This made it an extremely forgiving motorcycle, enabling a new line to be introduced, part way around a fast bend, or the throttle to be relaxed, even braking while cornering, should it be necessary, to be accomplished in utter safety. The only time the legendary transverse torque reaction was noticed was in the power band, when dropping the powerful clutch after a snappy gearchange with the throttle open; then a thrill ran through the 'Guzzi as it bit

Top: **Shaft drive of the 'Guzzi is as compact as a chain drive. The silencers are German type. USA pattern gives a slight increase in power and more noise.** Bottom: **Car type electric starter. The grilled panel covers the air cleaner and battery. Oil filling was near impossible, owing to the silly little filler. (Hexagon bolt head below the barrel.)**

into the new ratio. Far less than dangerous, it added a bit of spice to the ride. Hailwood on the wicked GP Honda all over again, I was. All in all, the roadholding was of the sort to give a clip-on and rear footrest merchant the shock of his life if he tried to outrun it. At ultra low speed, the balance was perfect; traffic threading simplicity itself. Once it did fall, though, the weight was awesome, but that only happened when I was careless. Unless a complete halt was called for, riding at the slowest pace was feet up.

I have just realised that unlike most machine test reports, until now I have not consulted the notes I made at the time of riding, about the Ambassador. Everything about this machine has burned itself on my mind, but even allowing for that, I should check back to Carmarthen Bay and the Bristol Channel (I made my notes on an Esso map) to ensure that all points have been covered. Because of my earlier rambling, without notes, some of this will be in recap form.

Nothing is rubber mounted, neither

engine, nor controls, nor seat, so any vibration had to reach me. It did, up to 3000 rpm. Vibration is the only word familiar enough with most riders to use, but in truth, the sensation was more gentle than that; just a rocking wobble, not unlike a Commando's, less pronounced, yet firmer. At tick-over speeds of 800/1000 rpm, nothing much at all, but as power was applied through increasing clutch pressure, and the revs built up, so the wobble's pitch raised until by 3000 rpm it reached maximum intensity, then by a handful of revs more, was gone completely. It can in no way be compared to the tremors of a vertical twin, and is only mentioned because of its existence, *not* because of any unpleasantness caused. Think of the Ambassador as a smooth, relaxing motorcycle and you have it.

Talking of the clutch reminds me to mention how smooth it was to activate. Oddly enough it felt very similar to the Laverda's, which is not surprising, because they are of similar design. Stiff when immobile, but, once the engine was turning, comfortable. Not really light, but about the same as most diaphragm, dry units. Take up was, like others of its type, more sudden than the old, wet, multi-plate things, but no less controllable; in fact even more so in traffic, for it never overheated, slipped, or grabbed.

Gear changing was, in all fairness, only reasonably good. I am sure I remember reading that the first of these engines' gearboxes were equipped with pre-selector mechanisms. Why on earth these have not is beyond me for the arrangement suggests that what is good enough for cars, is good enough for motorcycles. It would help greatly. As it is the change grates as the new gears mesh. Slow changing helps, but not entirely. The heel and toe rocking pedal was a boon in wet weather, simply because rain always penetrates the upper of my boot rubbed by the pedal rubber; besides which, downward foot movement is more comfortable, so if it can be carried out all the time, then I like it. Granted, slick racing changes by this method are not simple, but then this is no racing machine, and would suffer appreciably were racing techniques applied to it.

The frame is a full loop, tubular steel type; entirely conventional in geometry. The swinging arm pivot moves on two taper roller bearings.

Like all Italian machines, and most Japanese, the Ambassador's paint and chrome are only mediocre; not really bad, but not so good either. All the engine and transmission joints remained oiltight; their alloy castings were strong and superbly made. This goes for the complete machine. These models will undoubtedly last for years due to their immense strength and size, probably to Moto Guzzi's despair, for they will last too long, and require obsolete spares.

Twin carburetters are unnecessary, and not in my mind, particularly desirable on a machine of this character. They require synchronising too often for my liking. One carburetter on a twin cylinder machine is so much sweeter, and loses no measurable performance, but two it has, and I must admit, they seem well matched. Both breathe through an enormous air cleaner.

The battery is, like the Laverda's, huge, and the electrical system all the better for it, although I intensely disliked the lack of an ammeter, and the lights could do with some improvement at only 45 w mainbeam. The system and machine deserve a 60 w sealed beam or quartz halogen unit at least.

Like the Japanese, the Italians seem incapable of manufacturing good brake linings; these were poor. A few hard stops and the front brake weakened. Even on the first high speed deceleration I could feel them giving up in the front drum, as the lever closed on the handlebar. The brakes themselves seemed fine, without any hint of that awful tls servo slam, and with a tremendous stopping potential once properly linered. They remained watertight even after parking.

And that, gentlemen, is the end of a report that by rights deserves another four pages. Many of today's machines are designed to appeal to the playboy motorcyclist, as an adult toy. This the big 'Guzzi will never do. It's built for men who have within them the love of truly great machines, and an urge to travel. A long way. Fast.

Specification: Engine: Four-stroke tranverse 90° V twin. Capacity: 757.5 cc. Bore and stroke: 83 x 70 mm. Compression: 9:1. Bearings: all plain. Con-rods: steel. Carburetters: 2 x 29 mm Dell 'Orto VHB (Square). Lubrication: gearpump at 60 lb sq. in. Filter: wire gauze. Transmission: Gearbox; four-speed, positive stop, with down-for-up gear change by RH rocking pedal. Ratios: 11.62, 7.59, 5.73, 4.51 top. Transmission shock absorber contained with gearbox on clutch shaft, sprung-loaded, twin lobe face cam. Clutch: crankshaft-mounted, twin-plate, dry, diaphragm loaded. Electrics: Marelli Ignition: coil and distributor. Starting: electric. Generator: 12 v 300 w at 2400 rpm, belt-driven from front of crankshaft. Lighting: 45 x 40 w headlamp, containing fuse box. Frame, full loop, tubular steel. Suspension: front: fully damped telefork. Rear: load-adjustable, fully damped swinging arm. Brakes: Drum. Front: nine in. tls. Rear: nine in. sls. Wheels: both 4.00 x 18 in. WM4. Dimensions: Wheelbase: 57¾ in. Clearance; five in. Weight: claimed dry, 502 lbs. As tested with full fuel, oil, rear crash bars (front crash bars standard), panniers, windscreen and road-dirt, 563 lbs. Claimed petrol consumption; 44.2 mpg. (Test conditions unknown) Test consumption; 55 mpg under medium touring conditions. Fast touring not tested. Claimed top speed: 115 mph. As tested: 106. Concessionaire: Don Barratt Motors, 13 Station Rd., Redhill, Surrey. Tel. 71–61750. Price: £962. 1s. 6d. (Inc. import duty and tax).

DAVE MINTON

RETURN OF THE V-TWIN!

GREAT

Guts by name and guts by nature, that's the Guzzi shaft-drive 750 cc V-twin. And let's get it correct at the start—you say Guzzi as if it were spelt "guts-zee".

This is good because it so well sums up this 115 mph development of the original 700 cc version.

The 750, called the Ambassador in the USA, has an "over-square" motor with the crankshaft set along the length of the frame.

Output is a claimed 60 bhp at the high revs, for this class of engine, of 6500 rpm.

It's all-alloy, the compression ratio is 9:1 and the valve gear operated by push-rods. Despite the cylinder size and the compression, the Italian factory have managed to produce a thoroughly reliable electric starting system. So sure are they of this that no kick-starter is fitted!

The twin-plate dry clutch is on the back of the engine and the four-speed gearbox is bolted on car-style. There is an unusual gear primary reduction between the engine and gearbox, from which there was a very faint background whine at all times and in all gears.

The rest of the machine is fairly orthodox. Guzzi are not even the first people to put the right-hand side prop-shaft into the swinging fork.

Details

Detail-wise, there are some unusual features. The 29 mm Dell'Orto carburettors are of the flat-slide type. Cylinders are chromed alloy, that is, no iron sleeves and the crankshaft is one-piece with plain bearings.

Pistons, in common with many Italian engines, have four, not three rings, an extra scraper being placed at the bottom of the skirt.

This not only helps oil control but quietens the motor by restricting piston tilt as the con-rod moves through its angular range during the power stroke.

The electrical side is really massive by motorcycle standards.

The 12-volt starter churns out nearly three-quarters of a horse-power. The starter button is

GUZZI

Bruce Main-Smith rides and reports on this Italian heavyweight

Maintenance-wise the electrics, in the form of a distributor, carburation with twin Dell'Orto 29 mm units, and tappets are all easily reached, as can be seen above. Allen screws are used extensively on the motor

The twin-leading-shoe front brake is more than a match for the performance of the V750. Note alloy rims are standard units

No kick-starter, just this $\frac{3}{4}$ hp electric starter which proved foolproof throughout the test. Engine is wet sump, car-style

Shaft drive a la BMW type. Note large and very effective silencers, also twistgrip suspension leg adjuster—a model machine

positioned on the handlebar.

To charge the massive 28 lb. battery, the Minarelli DC generator produces 300 watts, almost three times the output of the alternator on any British big bike.

All this means you are sitting astride 550 lb. weight, not to mention £962 in cash!

We were very conscious of the sheer expense as well as the exclusiveness of the V750 when we took our sample out for test.

There is no doubt from the machine's behaviour, the Italian ex-racing factory's repute and the reports published in America, where the V750 is used by the Los Angeles Police Dept., that the claims for the V-twin are justified.

These are a top speed of 115 mph, standing-quarter of $14\frac{3}{4}$ sec. with an end-speed of a staggering 92 mph, and a fuel consumption of around 50 mpg.

Curious

The V750 is a curious machine to assess, for this reason:

It will produce all the results you usually want without the rider making any particular effort.

This is true of the engine, the gearbox and the handling.

Dr. Jekyll, so suave and unpretentious, can then be changed into a ravenous beast of a road-burner by cracking the whip and *driving* the thing.

Take "Dr. Jekyll" first.

Everything goes smoothly and easily. The engine, which has no dramatic cam effect or need to be kept on the bugle, responds well at, say, 45 mph in top to being tweaked.

Stay in top if you like and squirrel in and out of traffic. There's plenty of lusty, low-down power.

In a sustained burst on getting a clear stretch in these circumstances, the motor gets well into its stride at 3000 rpm, about 55 mph.

This surge persists all the way up to peak and the 6500 on the rev-counter comes up much faster and more smoothly than you expect.

A considerable amount of low-speed harshness, mainly shaft-drive snatch, might lead

ROAD IMPRESSIONS TEST

MOTO GUZZI V750

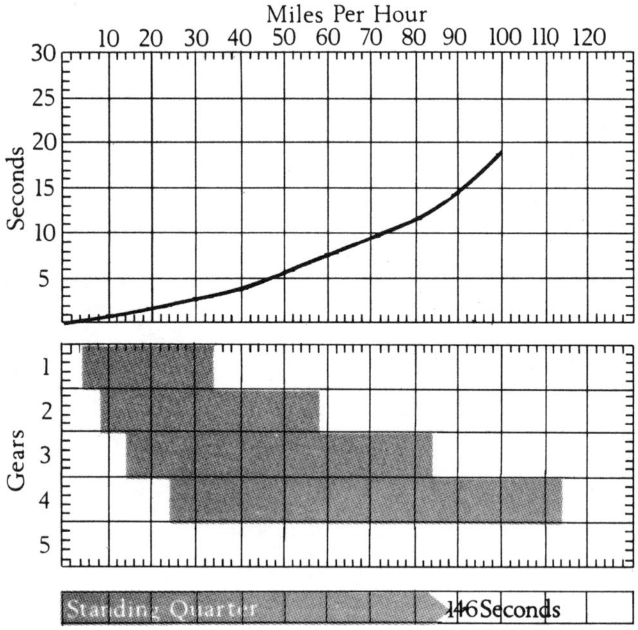

fuel consumption (average figure) **50 mpg**
braking from 30 mph **29 ft. 6 in.**

SPECIFICATION

Engine: Twin-cylinder, four-stroke. Air-cooled V-twin with crankshaft in-line with frame. Capacity 576 cc, bore × stroke 83 × 70 mm. Compression ratio 9:1. Power output 60 bhp at 6500 rpm. Lubrication by gear pump with oil in sump. Carburettors: twin 29 mm Dell'Orto VHB with air cleaner.

Transmission: Primary reduction by gear. Two-plate dry clutch. Four-speed footchange gearbox without kickstarter. Ratios overall: 1st 13.4, 2nd 8.0, 3rd 5.7, top 4.5:1.

Electrics: Minarelli 12-volt 300-watt dynamo with 32 amp. hr. battery. Electric starting. Battery and coil ignition.

Wheels: Pirelli 4.00 × 18 in. Super Sport tyres front and rear, with $8\frac{1}{2}$ in. brakes on both wheels, two-leading-shoe at front, single at rear.

Suspension: Telescopic front forks. Rear: swinging-arm with adjustable dampers and final-drive prop-shaft integral with swinging-arm.

Dimensions: Wheelbase 57.5 in.; ground clearance, 5.8 in.; dry weight 545 lb.; fuel tank 4.4 gal.; oil (in sump) 5 pt.

Price: £921 1s. 6d.

Importer: D. H. Barrett Motors, 13 Station Rd, Redhill, Surrey.

"Gutsy Guzzi"

one to expect vibration at high revs. Far from it. The V750 is very smooth indeed.

The exhaust is well modulated and there's no Honda-like "wailing" from the induction.

However, valve clatter is very apparent, presumably because the projecting cylinders don't have their heads shielded by the fuel tank.

If you want to unleash the full fury Guzzi, then simply wind it wide open. You can rev the motor like a Japanese buzz-box if you wish. The only drawback here is that the engine-speed clutch has a marked flywheel effect when you come to change up.

All gearchanges are best done in a leisurely manner.

You can boot them through, but you'll get an agricultural lot of clonks and an unmannerly leap forward as the clutch is dropped.

Do this after maximum revs in bottom and as second is loosed off, the front wheel lifts off the ground.

Bottom is a fairly sensible ratio, good for nearly 40 mph, about equal with the usual British 650.

Second's the same, more or less the mid-50s. But third's getting on for 90 mph.

This progressive gearing avoids unpleasant gaps.

Down-changes want attention but a little effort brings the reward of a completely silent gearchange.

The clutch, by the way, has stacks of bite, no suddenness and is no heavier than, say, a BSA Thunderbolt. Nor does it develop that nasty "slack" feeling after a full-gun burnout from the line.

So the gearing, as with the engine, you can treat two ways. Take it leisurely, or boot them through—the choice is yours . . . the Guzzi will deliver the goods either way.

Handling is also fair.

Doddle along and all's fine. You note the 550 lb. of the monster, of course, but only when you're manhandling it down the passage at the side of your house, or footing it in a stop-start traffic baulk. The rest of the time it just feels like any other 650 cc machine. Poddling about, you don't even notice the in-line crankshaft.

The V750 nails itself nicely into bends and sweeps out of them a treat. Then you crack the whip and "Mr. Hyde" has his say.

There is something of a tendency to understeer on right-hand bends.

You find either you come out slightly wide, that is near the gutter, or you compensate for this, from the apex on, by applying a little more "turn with the bars" than the angle of bank seems to suggest.

You only find you have to do this when you're having a bit of a "scratch".

Paradoxically, you do *not* find need to hold lock off when making left-hand turns. Here we are talking about sweeping curves at a reasonable speed, not feet-ups in the road to execute U-turns.

The reason why left-handers are not affected the same as right-hand bends is presumably a function of the natural camber of the road.

Suspension action is on the hard side, no argument about that. Most of the minor irregularities in the road are jobs for the dualseat to deal with, which it does very well.

You can see that the teles don't compress under minor bumps and feel that the rear end is of the same mind—the seat takes the shock.

This makes the suspension action a joy when you're hammering along at a high rate of knots over indifferent roads.

Braking power was impressive. The front unit is a two-leading-shoe device, well up to the bike's performance.

A curious tendency is to rely a lot on the back brake when riding the V750, possibly because of the flywheel effect.

The back brake is a good 'un and both brakes had that essential bite in the last few yards of a stop.

What of the machine generally, for it has an advanced specification?

It started easily, hot and cold. Idling was of the plonk-plonk-plonk type.

Carburation was clean.

Controls were well laid out and the electrical gear superb.

Inspection of used V750s, in for servicing and owned by D. H. Barrett's customers, showed them to be oil-tight.

In fact, there was so very little that one might be critical about. But, at almost £1000 one would hardly expect many faults . . .

MOTO-GUZZI V7 AMBASSADOR

IT'S A GOOD THING that tastes differ. Where, for example, would Moto-Guzzi be if everyone insisted on riding light, nimble motorcycles? Light and nimble the Ambassador ain't.

The big 750 is tank-like, in fact, and therein may lie a clue to its appeal. It looks bullet-proof, and that characteristic probably goes more than skin deep. In the course of a single working day at CYCLE WORLD, we receive a large pile of reader mail, much of which is devoted to moaning and groaning about the ills befalling a particular brand of motorcycle. Since the first 700-cc V-7 was introduced in 1967, and followed by the larger 750 in 1969, we have received no grave criticisms in the mail regarding this Italian V-Twin. Only praise. Even the local fuzz here and there are beginning to acquire Moto-Guzzis; considering the conservatism and pickiness of most police forces in regard to new equipment, this trend should tell you what they think about the Guzzi solidness and reliability. The influential LAPD has 85 of these machines in service, and even the California Highway Patrol has bought four of them for evaluation.

So what relevance does the Guzzi have in the civilian world? It is certainly not a sporting machine, in the sense that most American road riders have come to know. Rather, it is a heavy-duty tourer, with an Eldorado ride. It is half Harley (FLH) and half BMW, i.e., the kind of machine you would take on a short jaunt to Uruguay, or halfway around the world and back. Its heaviness would be an asset, for the tons of baggage and camping gear you'd take with you—mounted in the most unroadworthy places—would hardly be noticed.

The 1971 version of the Moto-Guzzi is not much different than the first 750. Refinement, much of it in cooperation with police department clients, has been the keynote in the evolution of this machine. Externally, the machine looks much the same, a deceptively low silhouette, a blend of big, upthrust cylinder barrels, candy apple red fuel tank and side panels and smart white enamel fenders, pinstriped in red and black. The most noticeable innovation outside is a new instrument layout. Instead of the single speedometer unit surrounded by a polished aluminum housing, there is now a group of separate Veglia speedometer and tachometer units, and four idiot lights indicating neutral, oil pressure, ammeter and lighting status. The ignition key is also in this grouping, and the keyhole is protected by a small, swinging coverplate. The "dashboard" for this grouping is a dull black, preferable to the previous housing, which bounced annoying reflections into the rider's eyes. The layout is neat and simple, so we are reluctant to complain about the fact that the idiot lights aren't labeled, and are all red, except for the yellow neutral light. And the tach needle, on this machine, was a wiggler.

Another barely noticeable change cleans up the side view of the motorcycle; the battery covers now extend to cover the air cleaner.

Internally, the most important modification occurs in the clutch throw-out bearing, which was of the caged ball variety and was predisposed to early failure. This is now replaced by a needle bearing. The clutch operating arm mounting is also strengthened by a bigger screw.

The Guzzi's engine could be described as a "perfect configuration" V-Twin. Its 90-degree cylinder angle permits 100 percent balancing to cancel primary vibration, at the small

Yes, Virginia, There Is An Alternative To The Electraglide Or The BMW.

cost of adding a minimal rocking couple, due to the fact that the connecting rods run slightly offset on the same crankpin. The elimination of primary imbalance is the reason that BMW's Twins have a 180-degree cylinder configuration. A slight advantage accrues to the choice of a 90-degree V in that it tucks the cylinders up and out of the way, making the engine package more compact. However, the firing sequence of two cylinders offset at 90 degrees is uneven. So the Guzzi, at engine speeds just above idle, jiggles with the staggered rhythm of a reducing belt machine at a ladies' health spa. Fortunately, for those of us who don't want to lose weight, these oscillations smooth out quickly above 1500 rpm.

As can be expected in any tranversely mounted engine, where the crankshaft is running parallel with the frame, the Guzzi manifests a torque reaction that pushes the machine to the side when you turn the throttle on or off. This is mainly felt when the machine is not in motion.

The reasons for mounting a 90-degree V-Twin transversely in the frame are actually three.

One, it fits better, particularly in the case of the Guzzi, which has a deep, finned wet sump crankcase.

Two, the most important, transverse mounting is peculiarly appropriate where shaft drive to the rear wheel is to be employed. The driveshaft rotates in line with the frame, and it is desirable that the crankshaft do so too, if the mechanical weaknesses and inefficiencies of changing rotational direction 90 degrees are to be avoided. By way of contrast, the 45-degree Harley Twin, a more compact "V," is mounted longitudinally in the frame and therefore drives the rear wheel with a chain. The new 750 Ducati, a 90-degree V layout, is mounted longitudinally and uses chain drive also.

Three, transverse mounting allows the Guzzi's frame to be kept low, benefiting the short rider, and at the same time keeping the cylinders out in the cooling airstream. The transverse mounting also scores well for easier maintenance, not to mention the aesthetic benefits of hiding the busy clutter of electrical components–generator, distributor, wires, etc.–behind those cylinder barrels.

The engine is basically set up to be a tractable "torquer." It displaces 757cc but is rated at 60 bhp at a modest 6500 rpm. The carburetors, a pair of Dellortos, are of moderate choke size, 29mm. Flywheel effect, produced by a massive crankshaft, flywheel, ring gear, etc., is quite pronounced, making for sedate throttle response, and a not-too-subtle leaping effect during rapid gearshifts. The flywheel effect, of course, is quite advantageous at low engine speed. It may be pottered through a 25-mph zone in high gear with incredible smoothness. Its sedateness should not be construed with slowness. This great Beastie will definitely move out, unless 86-mph quarters don't impress you.

Conservative engine design, automotive in character, is the keynote of the Guzzi. The crankcase, cylinders and cylinder head are in light alloy. Alloy pistons, with four rings, run against long wearing hard-chromed barrels. The overhead valves are pushrod operated, both pairs by a single camshaft in the crankcase cavity above the crankshaft.

Plain bearings are used to carry the one-piece forged crankshaft, as well as the connecting rod big and small ends. Accordingly, the oil lubrication system operates at a relatively high 54 to 60 psi, controlled by a pressure relief valve. The

crankcase is unusual in that it is of one-piece construction, which both increases strength and eliminates oil seams; the crankshaft may be removed easily from the rear, though, after removal of the clutch/transmission unit.

A heavy-duty double-disc dry clutch transmits power to the four-speed transmission. The clutch is virtually slip-proof, and the transmission, operated by a double-tang foot lever in a one-up-three-down pattern on the left, is smooth and sure-shifting. However, the heftiness of gearbox components makes for some noticeable clunkiness, especially during rapid upshifting, or during uncareful downshifts. The important thing is that, in spite of the unruly noises inherent to this unit, it always clunked *into* gear under any situation. It never bounced out or jumped into a false neutral. When the machine was at rest with the engine idling, it was quite easy to slip into the neutral position between first and second gears.

The automotive theme is carried into the electrical system. The bike is a push-button starter, by means of a 0.7 bhp, 12-V motor driving directly to the flywheel through the starter ring gear. There is, in fact, no auxiliary kick starter, nor need there be. The battery, which weighs about 28 lb., is rated at 32 amp/hours and the generator is a belt-driven 25-ampere, 300-watt Marelli, more than enough to drive the starter motor, lights and host of accessories necessary for police work.

The maintenance aspect of the bike is quite encouraging. A backyard mechanic can only smile at all the big bolts and hefty filler plugs, all easy to reach and promising immunity to mauling from the most gauchely handled economy size crescent wrenches.

Pleasant surprises are due when you first swing a leg over the Guzzi. Seat height is low. This, of course, means that you can firmly plant your feet on the ground when the machine is stopped. The seat itself is an orthopedic marvel. It is soft, well-shaped to support the thighs. Halfway along the seat, a step rises to support your tailbone. The passenger sits under similar circumstances. It is about the closest thing to dual bucket seats that a motorcycle can have—short of looking outrageous. The relationship of seat, footpegs and handlebars is comfortably compact. The bars are the typical American-style wide variety, but not too wide, nor too high. Thus the rider ends up leaning forward slightly, a good position at freeway speeds.

Suspension seems slightly stiffer than on the first 750 we tested, and the new machine didn't have as much tendency to heave up and down on acceleration and deceleration as did the old one. Front end damping seems improved. A small lack in rebound damping ability is indicated at the rear, for the Guzzi gallops somewhat in recovery from a briskly crossed rain gutter or severe "thankyoumam."

Cornering will be accomplished in comfort with the Guzzi, although not with the same elan that one associates with a sports-oriented roadster. For one thing, the front end is heavy feeling. The combination of front suspension, instrument panel and a headlight mounted way out in front is in fact a heavy one. Combine this with a wheelbase of nearly 58 in. and you have a slow steering motorcycle. The jerky effect of a heavy flywheel and the heavy weight of the machine do not lend themselves to "stuffing" the bike into a hairpin, or scraping the pegs in a faster bend. But within its limits, the Guzzi is solid feeling, imperturbable, and secure in its mission—to maintain a steady high cruising pace for hours or days on end.

Braking is just short of phenomenal, but quite a bit of hand lever travel, and pressure, is required before the front tire starts making nervous sounds. Yet, stopping distance figures reveal that the Guzzi is in the ballpark with the best big bore machines weighing 100 lb. less. The worst part of the stopping phase occurs in the 30-0 mph range. This is probably due to the bike's cumbersome feeling at slow speed. Controllability of a machine, or confidence therein, has much to do with the results you get in a braking test. It is no surprise, therefore, to discover that most bikes weighing less than 400 lb. do better (assuming adequate brakes) than the heavier ones, at least in the 30 to 0 stop. However, heavier bikes like the Guzzi have as good a chance to do as well as the lightweights in a 60 to 0 stop. We noticed some fade in repeated brisk stops from high speed as the front drum stored heat. However, the effect is not serious, and highly unlikely to be encountered in typical Guzzi riding. The rear brake is effective, although the foot lever is an inch too far from the footpeg for the most convenient control.

Highway cruising, as expected, is a calm, effort-free experience. The Guzzi transmits vibration to the rider only during heavy acceleration, particularly from low rpm, but is actually quite smooth as engine speed approaches the 6500-rpm redline. Both exhaust and mechanical noise level are low; the latter is quite surprising, considering the length of the pushrod valve gear and the exposed cylinders and heads. Part of the excellent mechanical quietness is due to the close tolerances and tight clearances used in this engine. The Moto Guzzi manual indicates just 0.0021-.0025 clearance for the pistons. That and the four-rings (two compression, two oil control) make for slap-free, quiet running.

What then is this creature, the Moto Guzzi? It is in a league with the FLH or the BMW, yet has its own individuality. In weight, it is in between the two. It has much of the character that pleases the riders of these other two brands. Solid comfort, longevity, calmness, heft. In spite of its slight unwieldiness below 20 mph, it scores as good or better than the other two in traffic. Its two-up behavior on the freeway is outstanding, and passenger comfort is excellent.

It will surely please the rider who likes to take his joys quietly, in great, gamboling doses.

CYCLE ROAD TEST

Moto Guzzi 750cc Ambassador

Another specialist in a world of single-purpose motorcycles, this one as focused as a motocross machine or a dragster.

Motorcycling in general has changed beyond belief during the past ten years, but the long-distance rider remains unaltered. He is usually quiet and unassuming; he often collects roadmaps, because faraway places hold a special kind of magic for him. He is forever afflicted with the urge to drift. Whenever he gets the chance, he packs his gear and climbs aboard his bike and heads for the other end of the continent, to taste the freedom and the loneliness of the open road. He is there just for the pleasure of seeing it all from his saddle.

Long-distance highway riding fell into disfavor in the early 1960s. People decided that it was too dangerous, or that it somehow came uncomfortably close to the then-popular stereotype of the motorcyclist as a badass. A lot of people went into trail-riding instead. Street riders started buying short-haul traffic-light racers with small gas tanks, skinny fenders, lots of chrome, tremendous acceleration, vicious vibes, and notorious unreliability. Only BMW and a few other manufacturers continued to offer pure touring models emphasizing riding comfort at speed, and reliability above all else. In the mid '60s, Berliner began importing the big Moto Guzzi from Italy, to try to grab a piece of the touring-bike market from BMW. It was then a very small market.

Things changed in the late '60s. *Then Came Bronson*, with Silliphant-style scripts and Michael Parks' Magical Motorcycle which could change in midair from a Sportster to a Sprint, revived a forgotten fact—long-distance riding *is* the essence of motorcycling. That point was brought home to many people through the misadventures of Peter Fonda in the highly symbolic (and highly unrealistic) motion picture, *Easy Rider*. During that same period, many of the standard traffic-light-racer street machines were undergoing an amazing metamorphosis. They improved so much in riding comfort and reliability that you could now ride one of them a couple hundred miles without it shaking loose the fillings in your teeth, or dribbling off a steady stream of nuts and bolts and other parts along the way. For these—or maybe other—reasons, many riders have begun to discover the enumerable joys of touring.

For the person who spends most of his time in the saddle riding around town, and takes trips only occasionally, most of the big street bikes made these days will give him what he wants. There is still no all-purpose road bike, although the magnificent Honda CB-750 comes pretty close. But most of the confirmed long-distance riders are primarily interested in maintenance-free operation, and then in riding comfort; those riders will probably want to stay with the special-purpose touring machines, of which our test bike—the 750cc Moto Guzzi—is a good example.

Most riders will agree that the looks of a motorcycle are an important part of its character; few people would buy a machine without knowing what it looked like. When you study a bike, its lines give it a unique personality that jumps out and hits you in the eye. We studied the Guzzi. It looked long, low, compact, heavy, and a trifle scary. But its personality didn't come across. Our motorcycle had the most colorful paint job available on a Guzzi: red tank, toolboxes, and air-cleaner cover; white fenders; and black forks. But somehow it didn't appear gaudy or ridiculous with all of those colors. Nothing about the machine stood out, except for the unique opposed V-cylinders; even the massive engine cases looked natural in that machine. Finally we gave it up. Then, toward the end of the test, our editor said, "It's an old-line motorcycle; except for the suspension, it looks just like a 1955 bike." We looked through a bunch of the back issues of *Cycle*, and he was right. Thinking about it some more, we finally realized that the Moto Guzzi is all of a piece; everything is so well-proportioned and conservatively shaped that no part of the machine obtrudes,

A lot of the Guzzi's appeal as a long-distance bike stems from its seat—the best.

except for the cylinders. And sure enough, most long-distance riders want conservatively-designed, gimmick-free motorcycles.

The Guzzi (pronounced "goot-zie") was designed to cover many miles with a minimum of maintenance. Probably as much thought was given to the types of materials to be used as to the shaping of those materials; a huge, one-piece crankshaft assembly, large-diameter plain bearings, and a high-volume lubrication system; all-alloy cylinders with hard-chromed bores, so that the piston and the cylinder will expand at the same rate as they get hot; a gear-and-shaft drive from the engine to the rear wheel, eliminating primary chains, rear chains, and rear-wheel adjustments.

The frame is simple but sturdy. A pair of cradle tubes drop from the steering head to the base of the engine, which cross-braces them. The Guzzi doesn't bother with a rear downtube; the cradle tubes simply angle upward again, to mount the shock absorbers. As they start upward, the cradle tubes grow a pair of huge bosses that hold the swingarm in tapered-roller bearings; the swingarm serves as a cross-brace. The huge top-tube, lavishly gusseted at the steering head, runs way back—almost to the shock absorbers—to a T-joint connecting it with the cradle tubes.

Upon close inspection of the motorcycle, we found that the castings were clean and sharp; the chrome plating looked hard, not greasy; and the paint didn't show any runs or orange-peel. We didn't see any hammer marks or screwdriver scratches on the bike, but then, we really hadn't expected to see any, for several reasons. First, from all we can learn (without having visited the factory), Guzzi seems to have their quality-control procedures set up to such an extent that everything gets tested and assembled carefully to begin with. Second, the factory spends a lot of time fussing with the completed motorcycle. They average 30 to 40 miles of test-riding to check everything out, which is quite a bit higher than the average. Third, they crate and ship the *fully-assembled* motorcycle to this country (with the handlebars removed). Most other factories don't do that; instead, they partially-disassemble the machine and put the parts into a crate, then the dealer's mechanic reassembles the motorcycle.

The send-it-in-one-piece philosophy, versus the pull-it-apart approach, can make a big difference to a rider who is interested in reliability. Usually, the factory disassembles the machines for crating because the distributor (or importer) asks for it. Because the collection of parts will fit into a smaller crate than will an assembled motorcycle, the shipping charges from the factory to the distributor, and from the distributor to the dealer, will be cheaper. From the distributor's point of view, the suggested-retail price of the bike will be cheaper; the dealer has to worry about explaining the high setup charges to his customer. The additional setup charges often cost the retail customer more than would the extra shipping charges for sending the bike fully-assembled.

But the real problem is that the factory has no way to learn whether or not its people are making mistakes in disassembling the machine, since it has already been tested. And at the other end of the line, many dealers use untrained help to do their setup work, rather than apply their overworked skilled mechanics to the task. Several distributors have commented privately that they are tracing the vast majority of their warranty problems to improper setup procedures on the part of the dealers' mechanics. By shipping their motorcycles fully assembled, Moto Guzzi and Berliner have succeeded in preserving the careful workmanship that went into the original construction of the bikes at the factory.

The front end of the Moto Guzzi looks uncomplicated. The forks and headlight are all black, except for an exposed chromed section on each slider leg, and the chromed headlight rim. The forks are massive—proportioned to the rest of the machine. A black-crackle instrument panel, which looks as if it were designed for the bike and professionally manufactured, hides the fork crown and bristles with instruments and indicator lights. Two extension bars sprouting from the back of the fork crown mount the handlebars. The bullet headlight has the right size and shape for the Guzzi.

The front and rear fenders are large, to keep the road-muck off the rider, however they are so gently valanced that they don't seem to be as large as they really are, and Guzzi didn't make the mistake of chrome-plating them. The large front fender brace *is* chromed; it is curved to match the leading edge of the fender, and the effect comes off well.

The machine has huge brakes. The wheel hubs, with center finning, are complicated only where they have to be and are otherwise very plain, and they look good. The bike carries 4.00 x 18 tires fore and aft; there isn't much space between the Borrani alloy rims and the huge hubs, so the spokes are

For a big bike, the Guzzi is surprisingly nimble and agile.

Driveshaft serves as left-side swinging arm, operates silently.

Transverse engine sounds a lot like a Harley-Davidson Vee-Twin.

Choke-less 29mm Dellorto carbs are accelerator-pump-equipped.

Bike's instrument panel locates key-switch in a practical place.

short, contributing to the heavy, compact look of the motorcycle.

The chromed exhaust pipes and mufflers have a gentle, graceful sweep—they would look good on any bike, and they are gorgeous on the mighty Guzzi.

The red gas tank, with white striping and chromed panels at the knee indentations, is conventionally teardrop-shaped; it is proportioned correctly in height, but it looks (and is) a bit too wide. Its saving grace is that it holds six gallons. The red toolboxes and air-cleaner cover are uninspired—they were designed to fit the spaces—but they do not detract from the looks of the bike.

The huge saddle is simply styled and unadorned, and it looks almost as comfortable as it really is.

One of the reasons why the motorcycle hangs together so well in looks is the absence of excess chrome. Aside from the shiny parts already mentioned, the only other chromed parts are the rocker-box covers, the crash bars (which protect the aluminum cylinders in the event of a spill), the grab bars, and the fully-exposed rear-shock springs.

When you get aboard the Guzzi, you learn that it really is just as low as it looks. The 18-inch wheels contribute to this, as do the slanted cylinders that don't have to sit upright beneath the top tube and the gas tank, and the trick frame. Although the saddle is plenty thick, the saddle height is only 31½ inches, which meant that our five-foot eight-inch test rider could sit in the saddle and get both feet on the ground when stopped.

The Guzzi's centerstand gave us about the average amount of hassle for a big bike, but the sidestand was so neat that we used it for parking the machine throughout the test. Using the toe of your boot, you can flip the sidestand up or down without any fumbling, and the stand will support the bike on any normal paved surface.

Moto Guzzi does not give you the option of starting the engine either electrically or by kicking; they left off the kickstarter, so either the electric starter works or you're out of luck. (The electric starter kept working for us.) Since few motorcycle designers have as yet designed a kickstart lever that works when you want it to and stays out of your way the rest of the time, we didn't miss it. Had the battery been weak, we might have felt otherwise.

To start the engine, you first flip up the waterproofing cover located on the instrument panel atop the fork crown, and insert the ignition key. Click it to the right, and many lights come on, none of which are labeled. If you have memorized the applicable portion of the owner's manual, you will recognize the red no-oil-pressure light and the red electricity-not-generated light and (if the headlight high beam is on) the red high-beam indicator. That leaves a large orange light, which is the gearbox-neutral indicator; the transmission must be in neutral before you can start the engine.

In order to get neutral you pull the clutch lever, which requires very little effort. The gearshift lever lives on the left side, and it has a heel-and-toe arrangement, which saves shoeleather if you can remember to use the heel part of the lever. We didn't, which meant up for low, with neutral between low and second.

With the neutral-light lit, the next step in the starting drill is to turn on the gas. There are two beautifully-machined, no-drip fuel taps with click stops. Turn on one of the taps, and leave the other one off as a reserve.

The 29-mm Dell Orto carbs do not have ticklers. Instead, you rotate the twistgrip one time to allow the accelerator pumps to do their thing. If the weather is cool, you choke the carburetors by pulling the air-lever control on the handlebars. Then you blip the starter button, and the engine starts running.

Whether by sheer accident or through careful planning on the part of the Guzzi design engineers, the exhaust note of the motorcycle sounds almost exactly like that of a

PHOTOGRAPHY: LARRY WILLETT

What the bike is all about: the freedom and the loneliness of the open road.

Harley 74—quiet but authoritative.

You can just climb aboard some bikes and bomb away into the sunset, but that is not the case with the Guzzi. Here everything happens with slow, stately grace, which turns out to be a Good Thing. But in a few instances, you have to change your riding habits to accommodate the bike. Our 135-lb test rider initially viewed the Guzzi with apprehension, figuring that such a heavy bike would probably be awkward for around-town riding, and that the bike would probably corner sloppily on rough pavement at speed. It took him three days to get used to the bike, after which he went tooling up and down the California coastline with gusto, and he would have welcomed the opportunity to ride the machine across the continent.

Getting the Guzzi's wheels rolling is no problem. The clutch pull is exceptionally light, and the clutch itself has a smooth, gradual engagement. The twistgrip just keeps on rotating and is relatively insensitive. Low gear is geared low. With that combination of factors, it is almost impossible to stall the engine when you're starting off.

As long as you don't try to hurry the gearbox, shifting up is a delight. The lever has a long throw, but the gearbox tells you when you've made the shift, and we never missed one (except into neutral) during the test. Because of the easy clutch pull and the positive

The lightweight rider can lean the Guzzi through turns without having to work at it.

shifting, riding in stop-and-go traffic is no problem. Too, the bike balances well and is light on its tires even at very slow speeds, so you can dawdle along at one mph in neutral, waiting for the light to change, without having to saw at the handlebars to stay upright.

The engine starts to come on at 3500 rpm. You can ride the bike at lower revs, but then it takes forever to build up a head of steam. We preferred to keep the tach needle somewhere between 4500 and 6500 (redline) rpm; things happen in a hurry when you twist the throttle in that rpm range.

The designers wanted to give the bike lots of acceleration, and they did so at the expense of the powerband. In its present state of tune, the Guzzi really needs a five-speed gearbox. Since that was apparently impractical, and the Guzzi is a touring machine, the factory left plenty of overlap in the higher gears, but they spaced first gear and second gear too far apart. At 30mph, you're almost redlined in low gear, and you're below the effective powerband in second. That's really awkward.

Downshifting the Guzzi properly takes practice. The secret is not to make any sudden shifts until you have mastered the technique. You put on the brakes and let the tach drop off to 4000 or so; pull the clutch; wind up the throttle and hold it there until the engine is revving nicely—which takes a few seconds longer than you'd think; then make the shift and dump the clutch again. It takes a few days to learn to match the speed of the engine with the speed of the gearbox shafts, either by reading the tach or listening to the engine. After you've learned the technique, you can downshift smoothly. Until then (if you're used to flicking the twistgrip in a downshift), what happens is that you receive a monumental jolt as the bike slows way down in a great big hurry and the rear tire screeches its protest, and the machine lurches sideways for a moment. After the third day of riding, we found ourselves downshifting the Guzzi smoothly, without thinking about it.

On the test bike, we had trouble shifting into neutral. If was impossible to tell by feel whether or not you were in neutral, and the neutral light came on too quickly. We discovered this by jabbing the shift lever, observing the orange light, coasting to a stop at a busy intersection, and dumping the clutch. The next thing we knew, we were in the middle of the intersection and cars were slamming on their brakes. We survived. Rather than interrupt the test to fix the bike (or get it fixed), we simply stopped trusting the orange light and thereafter fed out the clutch gingerly when we thought the bike was in neutral; sometimes it was and sometimes it wasn't.

The only other malfunction suffered by

Continued on next page

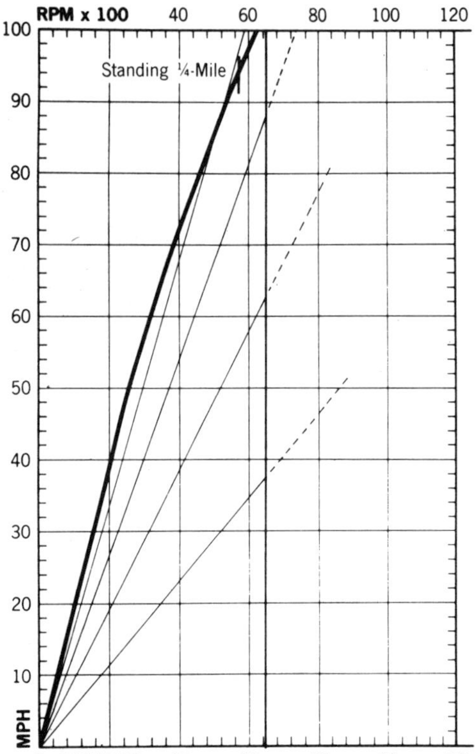

MOTO GUZZI 750 AMBASSADOR

Price, suggested retail	East Coast, POE $1679
Tire, front	4.00 in. x 18 in.
rear	4.00 in. x 18 in.
Brake, front	8.7 in. x 1.7 in
rear	8.7 in. x 1.7 in.
Brake swept area	90.1 sq. in.
Spacific brake loading	7.7 lb/sq. in., at test weight
Engine type	OHV 90-degrees V-Twin
Bore and stroke	3.27 in. x 2.77 in., 83mm x 70mm
Piston displacement	46.21 cu. in., 757.5cc
Compression ratio	9:1
Carburetion	(2) 29mm Dell'Orto
Air filtration	Paper element
Ignition	Battery and coil
Bhp @ rpm	60 @ 6500 rpm
Mph/1000 rpm, top gear	17.1
Fuel capacity	6.1 gal.
Oil capacity	6 pints
Lighting	12v, 300 watts
Battery	12v, 32ah
Gear ratios, overall	(1) 13.4 (2) 8.0 (3) 5.7 (4) 4.5
Wheelbase	57.8 in.
Seat height	31.5 in., with rider
Ground clearance	6 in., with rider
Curb weight	559 lbs., with ½-tank of gas
Test weight	694 lbs., with rider
Instruments	Tach; speedo, oil pressure, generator, neutral, and high-beam
Standing start ¼ mile	14.21 seconds 93.26 mph

the motorcycle (except when we brutalized it at the dragstrip) was the high-beam indicator light, which stayed lit all of the time. However, there was such a marked difference between the patterns thrown by the high beam and the low beam that we didn't need to depend upon the indicator light. The operation of the tachometer was marginal—the needle danced a bit too much, but we could read it.

In its around-town handling characteristics, the bike surprised us. We had thought that at low speeds it might be a bit awkward, but it proved to be reasonably nimble whenever the wheels were rolling. In La Jolla, there are signs marking out a bicycle tour among the narrow, twisty, ripply city streets along the waterfront, with lots of blind corners and sudden, unmarked intersections. We peeled off and dived into the bicycle-tour route, sailing along the lane-and-a-half streets with the Guzzi, brushing elbows with the occasional automobile, watching the wet-suited scuba divers trudging into the water, and finally playing tag with a Bonneville. Once you learn its preferences, the Guzzi makes an acceptable around-town motorcycle.

We took the test bike—while still new and tight—to the Orange County dragstrip to find out how fast it would go in the quarter. There we learned that the Guzzi breathes well. Although redlined at 6500, the engine revs easily to 7500 (by the tach), at which point the valves float. By preloading the shift lever, yanking the clutch with the throttle wide open, making the shift and dumping the clutch immediately thereafter, we could get into the 14s at over 90-mph. But the Guzzi wasn't built for that sort of thing. It would protest by leaping up and lurching sideways, and then it would recover. Several of us made many runs at the strip on the Guzzi. By so doing, we made the clutch a bit grabby, and with sustained overrevving, we split a short rubber tube to the oil breather. That wasn't a bad record; sometimes test bikes disgrace themselves terribly at the dragstrip.

The Moto Guzzi *is* built for long-distance touring, so we took it out on the open road to see how it would perform. For the first hour or so of our trip we forgot to wonder how well it was contending with the bumps, because it was doing such a magnificent job of smoothing them out. The suspension absorbs most of them, and it slows down the rest of them. The luxurious saddle is by far the most comfortable piece of motorcycle furniture we have ever sat astride—it alone more than justifies the existence of the Guzzi.

Our short test-rider found that the handlebars didn't have enough rise or rake to suit him, so he rode the leading edge of the saddle (which was quite satisfactory). Even so, the seating position was such that he could plant his feet on the pegs, rather than simply hanging them there as often happens. Thus he was able to use his feet to brace himself so as not to slide rearward on the bike while accelerating, or slide forward while braking. Our six-foot rider also found the seating position to be more comfortable.

The test bike didn't vibrate much. We could—if we thought about it—feel vibes in the handlebars, footpegs, and gas tank, but the vibrations were of a low frequency that we did not find disturbing.

The drivetrain had an expensive, cushiony feel—as if every component were mounted in rubber. Riding along at 70 mph (about 4500 rpm) in fourth gear, we could barely hear the engine running.

The steering geometry was just right. You could take one hand off the handlebars to make a signal, and the bike wouldn't flinch. Yet the steering was light enough so that even a lightweight rider could lean the machine through a whole series of S bends without having to work at it. You could also heel the bike over by applying reverse lock to the handlebars, but you didn't have to.

We took the machine along some narrow, twisty two-lane roads at speed to see what would happen. There we learned that somebody had done an excellent job of matching the suspension to the motorcycle, because the bike would hold a good line through corners on bumpy pavement. The wheels would stay on the asphalt, and not walk off the road.

On superhighways, we mixed it up with a bunch of semitrailers, and the machine remained rock-steady in the turbulence created by the big trucks. We could swoop past a large rig at 90 without feeling much more than a quiver.

Riding in the rain, we found that the great big fenders did an admirable job of protecting us from the roostertails thrown up by the tires. At night, we found the square pattern thrown by the low beam of the headlight interesting but not illuminating. However, the high beam did the job, and we could actually see where we were going at 80 mph. The instruments were easy to read, day or night.

The brakes were deceptive. They slowed the bike far quicker than they seemed. We used them vigorously and extensively at the dragstrip without fading them. We used them hard several times on the highway unexpectedly, and they did the job. We finished the test divided in our opinions of the brakes—one of us thought they ought to come on stronger; the other was satisfied with their operation.

The big thing about the Moto Guzzi is that it will charge along steadily, hour after hour, eating up the miles and carrying its rider in great comfort. It has been developing a mystique of its own, both among riders and dealers, over the past few years. It is a good motorcycle for the guy who wants to see what's going on at the other end of these United States—the true long-distance rider. ◉

CYCLE GUIDE

TEST REPORT

MOTO GUZZI V-750 AMBASSADOR

GUTSY GUZZI

The new V 750 Moto Guzzi tourer has impeccable road manners, cadillac comfort and the eerie smoothness of a turbine.

There are few very good specialized motorcycles made, but the Italian Moto Guzzi V 750 is one of these.

It is designed, built and sold with the tourer in mind. For pure speed, it's a no-compromise machine. Anything other than touring is a secondary consideration with the Guzzi.

There are only two other production machines that compete with the Guzzi, the BMW and the Harley Seventy-four. Other motorcycles, particularly sporting machines, are frequently converted for touring, but aren't really in the same field.

This is our first opportunity to road-test a Moto Guzzi. We found it a most interesting experience. Our initial reaction was awe at the Guzzi's huge engine. But first impressions can be deceiving. The Guzzi is big compared to the lightweight or middleweight sporting machine. But the V 750 isn't in this class. As a tourer it is average size.

While never intended to be a featherweight, the Guzzi weighs in at about 585 pounds, topped off and ready to roll. We found this out the first time we mounted the machine.

The Guzzi engineers have purposefully designed the V 750 to be muscular and beefy. They succeeded. The engine and driveline are so massive it appears as if the chassis was designed around the power plant. There is no place on the engine to fit anything larger than a pencil.

In many ways, the V-750's engine design conforms more to automobile standards than to those of motorcycles. Out testing revealed none of these detract from the handling and

One cannot consider the Moto Guzzi pretty, but it is a handsome and stout machine. Accessory windshield, fairing and saddlebag combinations are available for the tourer.

cruising qualities of the Guzzi

After adjusting to the Guzzi's sensation of massiveness, one fully appreciates its tank-like appearance. Actually, the Moto Guzzi is very handsome. Although every detail on the machine has been designed with touring in mind, the appearance of the package suggests the influence of Italian design. The huge engine is almost out-done by the monstrous 6.1-gallon fuel tank. We were most impressed with Guzzi's obvious meticulous attention to quality. We have never tested a machine with finer castings, painting or overall finish.

The Guzzi's forte is burning up the highway. In town and at slow speeds, it's awkward and clumsy; resembling a buffalo grazing on an egg ranch. This is compounded by a very long wheelbase, grumbling idle and a recalcitrant gearbox. The long wheelbase keeps the machine going straight on those lengthy stretches of highway. The noisy idle is created by the Guzzi's 90 degree V twin. The transmission operation and the gear ratios are not designed for hustling from one stop light to another.

The chassis is a very large, tubular, double-cradle frame. Like all shaft-driven machines, the swing arm serves as a combination shaft cover and rear fork. The right side is the shaft cover and the left, a conventional swing arm tube. Unlike other chassis, the frame members surrounding the engine flex considerably.

The mammoth 90° V twin engine is rough at an idle, but glass smooth at cruising speeds. Excepting the gentle tappet noise, it's almost impossible to hear the engine running at highway speeds.

The square slide Del Orto carburetors performed perfectly. They come with accelerator pumps to ease starting and to improve throttle response. The ignition is automotive oriented.

In most cases this is undesirable, but the Guzzi engine puts an enormous amount of side strain on the frame when the engine is revved. Earlier records reveal the Guzzi frame is one of the strongest in the business. The 58-inch wheelbase was created by the head stock on the domestic Italian model being pushed out several inches for riding most American highways. This has proven perfect.

The Moto Guzzi engine is the only mass production opposed (to the chassis) 90 degree V twin available. The BMW has a 180 degree opposed twin and the Seventy-four uses a 45 degree inline V twin showing three completely different schools of thought. The Guzzi V twin uses a large one-piece, cast crankshaft with both con rods revolving on a common journal. This reciprocating weight is offset by two large counterweights. All engine castings are aluminum and have much finning on the crankcase sump area, the barrels and cylinder heads. The cylinders also lack steel liners, using a hard chrome bore instead. This yields optimum cooling qualities.

The super long brake lever requires little pressure to lock up the back tire. The drive shaft and U joints are completely enclosed. It's much cleaner and quieter than a chain.

The toe-and-heel shift lever connects up to a great deal of linkage. When worn, these tend to make shifting more difficult. The huge electric starter gets the fires burning each time it's asked to with little complaint.

The gearbox is enclosed in a separate casting bolted to the engine. The torsional power from the engine is driven through the constant-mesh, four-speed transmission by an automotive type flywheel and a double-disc dry clutch. From the transmission, the power flows to the rear wheel via a double-U jointed enclosed drive-shaft. The car-oriented ring and pinion helical gear drive are in the rear hub.

Combining all these elements involved extensive design and testing for the Italian factory. On the highway you'll agree the fantastic results were worth the effort, but in town, these features are not all desirable. The engine's torquing action moves vertically to the chassis. Most conventional machines have this reciprocating force parallel to the frame. This, combined with the centrifugal effort of the clutch / flywheel assembly, produce excessive side force when the engine is revved. This is the same thing that happens when an automobile engine is revved. For the rider, this means a readjustment. At a stop or just accelerating, you must be prepared for that side thrust. Once underway, most of this side torsion disappears.

The Guzzi begins to kick up her heels at 30 or 35 mph. The bike's slow handling makes it feel clumsy at low speeds, but given the chance to unwind, the V 750 reacts like a well-trained thoroughbred entering the back stretch. At 3,000 rpm, the engine smoothes out like glass.

The power range extends from about 3,500 to 7,000 rpm. In this speed range, the engine's smoothness is absolutely uncanny.

The V 750 Guzzi is unquestionably the smoothest running machine at highway speeds we have ridden. The fact that the handlebars are bolted solidly to the fork crown and not insulated by rubber hints at this. The vibration at the handgrips is negligible. The only vibration we noticed was at the foot pegs, and hardly worth mentioning.

The Guzzi really performs best on the open road. The big V twin, 60 hp engine generates sufficient power to handle freeway speeds. It's astonishing how quickly and smoothly a bike this large reaches 60 or 70. The throttle responds admirably for accelerating or passing up to the century mark. Even without a windshield or

The huge (swept area) double leading shoe front brake provides more than adequate stopping power. Fading was minimal. The Pirelli tires are very roadworthy. Front forks are sturdy and give a superb ride.

The headpipes turned blue in a few hundred miles. We found that most premium grades of gasoline (mostly low lead types) resulted in occasional preignition. Retarding the timing a bit would solve this. Gas mileage is phenomenal at 55 mpg.

chine, it's hard to believe just how much noise a chain-driven machine makes, especially a big touring bike. The Guzzi's quietness at highway speeds is exceeded only by the BMW. The Guzzi's only noise is a constant mechanical tappet clash. If you like these mechanical noises, you'll not be bothered. We liked it.

The V 750 comes equipped with an infallible electric starter which sounds similar to an old Chevy six. To insure that the starter always works, a huge, 12-volt, 32-amp hour battery is standard equipment.

The carburetion is handled by a pair of 29mm 'square slide' Del Ortos. To further aid starting and throttle response, the carburetors are equipped with acceleration pumps.

Before starting the engine, we found it necessary to give the throttle

The handsome instruments are easily read day or night and free from any vibration. The idiot lights are for neutral, oil, generator and turn indicators (optional). The speedo is very accurate, but the tach lags a bit.

The cushioned hand grips are comfortable. The starter button is easily reached and brings instant action from the engine. The top lever is the choke. The gas cap doesn't leak a drop, but is clumsy to open and close.

The extra long fender keeps the water and road grime off the rider. The mufflers cut the exhaust noise down to a whisper, although they will rap some at high engine speeds.

fairing, cruising at 70 to 75 becomes a joy.

The Guzzi cannot readily or willingly be thrown into sharp corners. The handling is slow and prevents the rider from easily manipulating tight, twisty roads. On highways, with their sweeping turns and long straights, the Guzzi behaves impeccably. At 80, the rider could take both hands off the bars without experiencing any turbulent rèaction from the chassis. The faster we went, the more stable the machine became. Even with two up, high-speed handling was excellent.

The V 750's appointments must be classed as plush. The long dual saddle is perhaps the most comfortable seat we have tested. After several two-hour jaunts, we can pleasurably say that we never became saddle-sore.

• Many will wonder what is the advantage of using the shaft drive over conventional chain and sprocket. Besides the obvious elimination of chain adjusting and maintenance, driveline noise is completely eliminated. Unless you have ridden a shaft-drive ma-

several blips to supply the combustion area with gas. Rolling the throttle with the engine dead, though, will produce instant flooding. Our only objection to the induction system was a very audible drone when the throttle was turned full on, especially in top gear. Further enclosing or baffling the induction system could alleviate this.

One unusual Guzzi feature is the belt-driven generator. Though uncommon for a motorcycle, it performed flawlessly. The generator itself is also automotive-oriented, having an enormous potential output of 300 watts at 12 volts.

We were pleasantly surprised the first time we pulled in for gas. That huge gas tank took less than four gallons to fill. We figured the mileage at a phenomenal 56.5 miles per gallon. One must with the V 750 is the use of premium high-test gasoline. This is particularly important in larger cities where only new low-lead, anti-smog fuels are available. Even with the high test, though, the engine will occasionally ping when under a load. Retarding the timing might arrest some pre-ignition.

The V 750's instrumentation is very attractive, and easily read. The tach lags during acceleration, but the speedometer is the most accurate we have ever seen. Included in the instrument cluster are four idiot lights. One indicates neutral, two show low oil pressure and electrical charging and the other is available for turn signals. All are a nice convenience.

Moto Guzzi has created an impressive touring machine. It may well be the noblest Roman of them all. —*Dave Holeman* ***CG***

MOTO GUZZI V-750 AMBASSADOR

ENGINE

Type	90° V twin cylinder four stroke
Bore and stroke	83×70 mm
Displacement	757.5cc
Compression Ratio	9.0:1
Rated Max. Horsepower	60 @ 6,500
Ignition	generator/battery
Carburetion	twin 29mm Del Orto (square slide) concentrics w/accelerator pumps
Lubrication	pressure fed wet sump

DIMENSIONS

Length	88.3 in.
Seat height	32 in.
Wheelbase	57.8 in.
Ground Clearance	5.9 in.
Dry weight	557

WHEELS AND BRAKES

Front tire size	4.00×18 in.
Front brake type	double leading shoe, internal expanding
Rear tire size	4.00×18 in.
Rear brake type	internal expanding

TRANSMISSION

Type	constant mesh 4 speed
Clutch	dry, dual plate
Internal gear ratios	1st 2.230:1, 2nd 1.333:1, 3rd 0.954:1, 4th 0.750:1
Final ratio	4.375:1
Countershaft gear	8T
Rear wheel gear	35T

PERFORMANCE

Indicated highest one-way speed	115 mph
Acceleration 0-60	6.2 sec.
Quarter-mile acceleration:	
Top speed	92.17
Elapsed time	14:33 sec.

GENERAL

Air Filtration	dry paper
Battery type	12V-32AH

CAPACITIES

Fuel tank	5.84 gal.
Fuel reserve	1 gal.
Engine sump	3.25 qts.
Fuel consumption	56.5 mpg

FRAME AND SUSPENSION

Front suspension	telescopic double damping
Rear suspension	adjustable spring over shock
Frame type	tubular double cradle

OPTIONAL EXTRAS

Wixom fairing/windshield, saddle bags, turn indicators

STANDARD EQUIPMENT

Electric starter, belt driven 12V-300 watt generator

COLORS

Black, white & red

PRICE AS TESTED

$1,694.00 F.O.B. Glendale, Calif.

DISTRIBUTOR

Premier Motor Corp.
Hasbrouck Heights, New Jersey 07604

The four-leading-shoe front brake is superb. It is smooth in operation with no trace of fierceness when used at low or high speeds

An "easy start" lever is fitted to the carbs to make starting instantaneous when cold. The V-twin has wet sump lubrication

The car-type starter operates direct on to the flywheel, which is toothed on its outer edge. A kick-start is not fitted to bike

A QD rear wheel is made more so by this flap on the rear guard. Note wheel rims are alloy at the front and rear. Saddle is deep

IT'S A GREAT GUZZI!

Charles Deane reports on the most expensive "Superbike" in Britain . . . the Moto-Guzzi Sport 750

How many times do you get stopped by the law in a week? If you ride a Moto-Guzzi V750 Sport, then you can reckon on being waved to a halt about five times!

The first time, you get that "butterflies in the stomach, what have I done now?" feeling; the second time, "here we go again, I wasn't speeding" laced with doubt and then you get used to the numerous questions about capacity, performance, handling, braking, etc, from the enthusiastic motorcycle patrolmen.

Motorists alongside you at traffic lights wind down windows and make swift remarks like: "That's a beauty. What'll she

Shaft drive to the rear wheel keeps the bike a lot cleaner. Suspension units are adjustable for load. Silencing is excellent

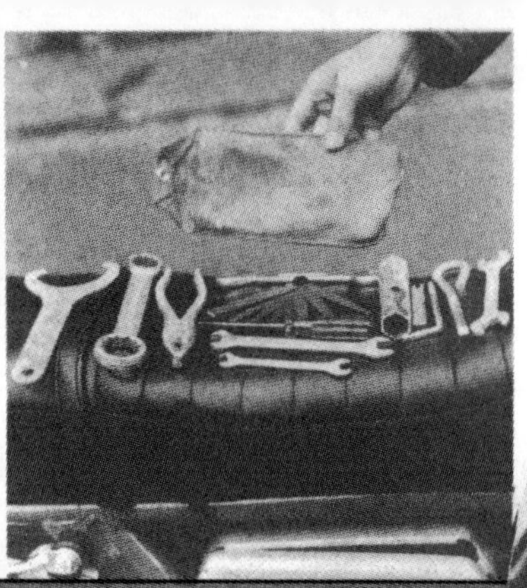

The toolkit is very good and suitable for most minor maintenance jobs. It fits into a lockable toolbox. Spanners are metric

The epitome of luxury! The V7 Sport motor looks as great as it is, producing 70 bhp at 7000 rpm. Note the deeply-finned sump with alternator housing above. Twin horns are standard

do?" or "Haven't seen one of those before. How much are they?"

Of all the machines I have ridden over the past four years, the Moto-Guzzi 750 Sport must be the prettiest, one of the most potent lumps of metal to tantalise the eyes of enthusiast and layman alike.

Moto-Guzzi appear to have taken one BMW, studied it very closely and produced a machine which is not only more expensive, but more powerful, faster on top speed and acceleration, better on braking and almost as prestigious in the eyes of the average two-wheel fanatic.

Moto-Guzzi earned their reputation with the famous single-cylinder racer in the hands of Fergus Anderson, plus the absolutely incredible V8 water-cooled 500 machine, which if developed might well have blown MV Agusta and Gilera into the weeds.

However, it was not to be and Moto-Guzzi withdrew from the Grand Prix racing scene to continue steady production of their horizontal single-cylinder machines in both two- and four-stroke designs.

Then came the superbike era, a demand from the U.S.A. for a large capacity motorcycle which would match the large pockets of this booming motorcycle market. The engineers at Guzzi had a deep thinks session and obviously agreed with the BMW way of mind and that is, if a thing is worth doing well, then let's do it properly. No messy chain . . . must be shaft drive! Who wants to kick-start a motorcycle these days? . . . Electric starting is better for the motorist who wants a bike!

Anyway, the outcome was the heavy, cumbersome but luxurious V-twin Moto-Guzzi V7, a bike ideally suited to the American "superbike" traditional market. But market patterns change and as Harley-Davidson quickly found out, trail bikes or sporting roadsters are really the scene in the U.S.A. today and Guzzi too have "switched on" with this new Sport V750, which bears little resemblance to the original V7 apart from the lateral 90 degree V-twin motor.

Power output has been uprated to a healthy 70 bhp at 7000 rpm and the four-speed gearbox has been improved to a five ratio unit.

Ignition system has been modified to a twin coil, twin contact breaker unit in place of the single coil distributor system of the V7 and the dynamo has been replaced by the high output alternator.

With a maximum of around 130 mph from the Sport V750, it is little wonder that Guzzi have had to improve the braking on this massive bike. A four-leading-shoe front brake replaces the original twin-leading-shoe unit, which is transferred to the rear wheel as a boost for back-end stopping.

Expensive

The entire machine oozes an expensive look. Wheel rims are alloy, mudguards are stainless steel, all electrical circuits are independently fused, steering is damped, not by a mundane friction unit, but hydraulically. The colour scheme of lime green and pillar-box red is dazzling and adds lustre to the sporty appearance of the machine.

That then is a brief description of the Moto-Guzzi Sport 750, but what's it like to ride?

Once again, it's BMW with a little bit extra . . . it has that bit more acceleration, a little more top speed and slightly better braking.

There is no kick-start on the Guzzi and it relies on a large 32 amp-hour battery to provide current to the car-type starter, which operates directly on to the starter ring on the flywheel.

There are two methods of operating the starter; the first is to switch on the ignition and continue turning the key, and the second is a starter button alongside the twistgrip throttle.

Each of the carburettors has a cold starting lever, and when used cold starting is instantaneous. The ignition key also automatically operates a steering lock when it is removed.

After allowing a few minutes for warming up, the cold start levers may be returned for normal running and first gear engaged with a slight click. If the throttle is blipped, similar to the BMW, the torque reaction of the "across-the-frame" twin causes the machine to twitch. This is more noticeable when stationary or travelling slowly.

The clutch is light and the take-up of the twin-plate unit is very smooth. Up to about 3000 rpm one can literally feel every power stroke of the V-twin, but once the rev counter starts creeping towards 4000 and from then on, the motor is remarkably smooth.

Silencing of both induction and exhaust is excellent and when the throttle is opened wide the only noise one really notices is the peculiar "chuffing" induction sound common to V-twins.

Acceleration doesn't seem all that vivid if one lets the clutch out fully before 2500 rpm is on the rev counter; but if given "the stick" and the clutch is dropped at 4000 rpm plus, the Guzzi gets away like a scalded cat. Then, if the revs are taken to 7000 maximum and the next gear engaged without letting the revs die, the punch in the back as you let the clutch out again is tremendous and almost pulls your hands off the bars.

The gearchange, similar to the BMW, is clonky under these conditions but if more sedate use is made of the throttle, a clonk is reduced to a click.

Ratios

The gear ratios are quite widely spaced and about town it is only possible to use the first three ratios to keep below the legal 30 mph limit. The motor doesn't run happily under 2500 rpm.

Fifth gear is virtually an overdrive for long distance, high-speed cruising. In fact, we had the Guzzi from new and with instructions from the manufacturers not to exceed 5000 rpm for the first 500 miles; it was found we could cruise at 70 mph on the motorway at 3800 rpm!

The manufacturers quote fuel consumption as 33 to 35 mpg, but with a 70 mph maximum on

ROAD TEST MOTO GUZZI 750

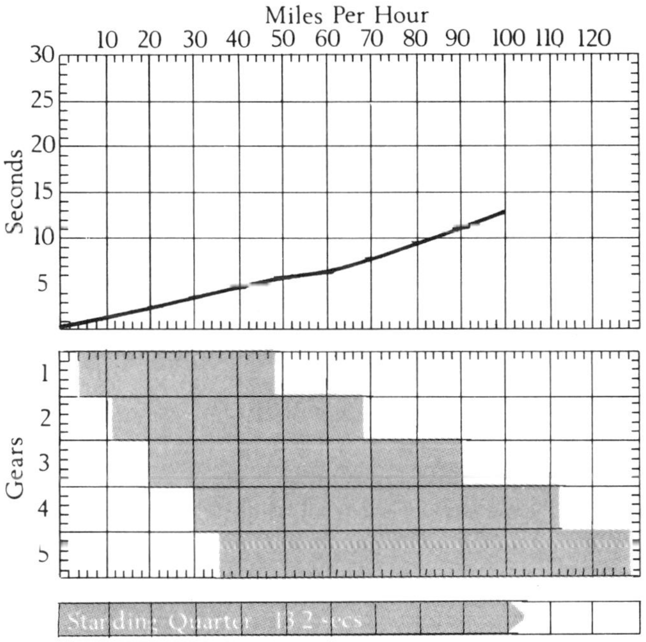

SPECIFICATION

Engine: ohv, transverse 90° V twin, bore × stroke 82.5 × 70.2 mm, capacity 748.39 cm^3, compression ratio 9.8:1, rated at 72 bhp at 7000 rpm. Two Dell'Orto 30 mm carburettors. Wet sump lubrication, 6 pints capacity.

Transmission: Five-speed gearbox driven by crankshaft-mounted twin-plate clutch. Final drive by shaft and crown wheel and pinion. Overall ratios, 10.828, 7.598, 5.739, 4.758, and 4.090.

Electrics: 12-volt AC/DC lighting coil ignition, supplied from 32 amp/hour battery and alternator. Starter motor, inspection light, independently fused circuits, flashers may be fitted.

Brakes: Front, four-leading-shoe, 220 mm; rear, twin-leading-shoe, 220 mm, cable operation.

Wheels: Front, 3.25 × 18, WM2 rim; rear 3.50 × 18, WM3 rim.

Suspension: Telescopic front forks, rear swinging arm with adjustable damper units.

Dimensions: Wheelbase 57.9 in., length 85.23 in., width 27.55 in., ground clearance 11.02 in., seat height 30.31 in., dry, fuel tank 4.4 gallons, including .5 gal. reserve, weight 454 lb.

Price: £1350 including P.T.

Distributor: Rivett's, 236 High St, Leytonstone, London, E.11.

"Super Superbike"

British roads, we accomplished about 45 to 48 mpg. It certainly makes a change to exceed the manufacturer's claims, although we later found that consumption dropped rapidly to the 35 mpg mark when larger throttle openings were used for acceleration tests and about-town work.

Handling at all speeds on the Guzzi Sport was reasonable, but there seemed a lack of damping at both front and rear which created a disturbing front fork action when cranked over on a bend. There was no twitching in the frame, simply a movement at the front which made it feel as though the front wheel was losing adhesion.

Maybe it was the thick-ribbed front tyre . . . it would have been interesting to have fitted Dunlop TT100 tyres to see if they improved the steering.

The one thing we did like was the low centre of gravity on the bike and it could be switched from left to right and back easily and swiftly. In fact, the Guzzi is a remarkably low-slung mount, for although I am only 5 ft. 6 in. short, I was able to put both feet firmly on the ground when astride the bike. I suppose this would make the riding position a little cramped for the taller rider, but for me it was perfect.

The dual-seat was beautifully soft and well-shaped, but was one of the weak points on this luxury mount. It absorbed water through the stitching like a sponge and for days after riding in a downpour we still had to use overtrousers because water continually squeezed out to soak the crutch of the rider.

The control layout of the Sport 750 was conventional, with the gearchange lever (one up and four down) on the right and the rear brake on the left.

The handlebar controls and instruments comprise, on the left, the clutch and full lighting console which operates the pilot light, headlight, dip and main beam, plus horn button. It was easy to turn lights on and off from the switch, but the horn button was hard to reach and operate and often one pressed the dipswitch button instead of the horn as both were next to each other.

On the right handlebar are the throttle twistgrip, front brake and auxiliary starter button, while in the centre are the rev counter and speedometer, plus a very good warning light system.

Oil pressure, generator, lighting circuit warning and neutral indicator lights are all in one console, but surprisingly there is no main beam warning light. Also, the neutral warning light occasionally gave false indication and, in spite of indicating that the bike was in neutral, when one released the clutch the engine stalled because the machine was still in gear.

The one vital factor which hasn't yet been mentioned is the braking and this cannot be faulted. A giant four-leading-shoe front brake, which makes one a little apprehensive about wet weather riding, was found to be smooth and gentle in operation. There is no trace of fierceness and stopping power is equal to the amount of pressure applied at the lever.

Confidence

The rear twin-leading-shoe brake is also superb and once again there was no sudden bite. Both could be operated with complete confidence in wet or dry conditions and from any speed without locking wheels or fading with hard use.

The one question which was repeated often was, "Is it really worth £1350?"

It's terribly difficult to answer for there is no doubt that apart from a few minor detail faults, the Moto-Guzzi V750 Sport is a superb machine in both design and engineering.

Mechanically it is reasonably quiet, except that one hears a little more tappet noise than normal because of the projecting cylinders and the shaft drive causes a whine on the overrun.

But you certainly get one hell of a lot of motorcycle for your money and it must be classified as almost unique and very select. It's perfect for long-distance, high-speed touring, doesn't jib at the occasional trip around town and for the ultra-enthusiast who has this sort of cash available to spend on his hobby, then it's a good machine to have in your stable.

By the way, Rivett's of Leytonstone have just become concessionaires for the Moto-Guzzi machines and will soon be importing other models in this Italian range.

ROAD TEST

MOTO-GUZZI 750 V7 AND SIDECAR

Splendid bike: but you can see why the sidecar outfit has died!

Guzzi and Astralite: "Not the most beautiful chair ever produced"

THE sidecar outfit. Outdated as the open-top bus . . . pipe-smoking dad and a load of kids . . . as lethal as an unbroken stallion . . . the most exciting form of travel on the road. How do you see it? At any rate, you'll agree that to come into the last-named category the sidecar should be something like a Steib or Watsonian Monza. Hitch on one of those monstrous saloon creations and no one loves the result except, presumably, the aforementioned dad and kids.

When I was told we had a Moto-Guzzi V7 and sidecar coming I took it as great news. I love sidecarring. It is just as well that I have one of my own for occasional use, though, for readers will not be unaware that three-wheeler tests within these pages are few and far between. In fact, there has not been one for over three years. This is not due to any particular foible of ours; there just aren't sidecar outfits around these days. New sidecars are virtually extinct, and very expensive. With the advent of the Mini only a diehard would choose to subject his family to the rigours of sidecarring when he can achieve the same result in more comfort and more cheaply.

"What's the sidecar?" I asked the Editor. "A Palma, I think," he answered, doubtfully. He was keeping something from me. Who cares, I thought, the chance to get to know the big Moto-Guzzi was too good to miss, whatever the sidecar. They can tack a bus on the side if they like. They had! Well . . . investigation showed it to be an Astralite, née Busmar. Not the most beautiful chair ever produced. Our model was originally a child-adult but the child's space in the rear had been covered to provide a sumptuous boot. Loss of the child's space had effectively reduced the passenger-carrying capacity of the Astralite to that of a sports chair, with the benefit of greatly increased luggage room and an almost embarrassing surplus of space for the passenger to enjoy. This in itself was not, to my particular passenger, much of an advantage. My wife, with far from fond memories of our Island holiday a couple of years ago, when we had a giant Watsonian on the side of my BMW, expressed a strong preference for the restrictions of a sports sidecar, complaining, rather graphically I thought, that she was bobbing about like a cork in the ocean in the Astralite.

All of which led us to conclude that the choice of sidecar to fit to the exciting 750 Guzzi was, well, strange. Like fitting a dray to a Derby runner. There were two reasons, we found: (1) A crashed Astralite handy in the Barretts of Redhill workshops (Barretts are concessionaires for the Moto-Guzzi and loaned us the test model) and (2) a dedicated sidecar man in the person of the manager.

We have ignored the Guzzi up to now. Not because we think it the secondary item in the outfit but because we felt readers would be eager for some explanation of this unlikely partnership.

The Moto-Guzzi Ambassador is one of the great unseens of the British motorcycle scene. You could count on one hand the number of such machines on the road in this country today; practically the only time the enthusiast will see one is in the Isle of Man at TT time. The high price is, not unnaturally, the prime reason why one so rarely sees the majestic V7. At around £950 the Guzzi is very much a machine for the connoisseur. But this does not wholely explain its rarety. Other £1,000 machines are to be seen with increasing regularity so, provided it is available, surely the V7 should sell? The key seems to be in the publicity afforded it. One never sees an advertisement for the Guzzi in British publications, rarely a road-test. There is no publicity agency extolling its virtues to Press and public; and so perhaps the prospective buyer is nervous of a machine that is so uncommon. Will spares be *always* available, will the local garage throw a fit if asked to repaint it? There are many good reasons for not buying an exotic machine, only one for buying one. Wanting it. The object of our test is to try and help the reader decide if this is the sort of machine he does want, which includes the all-important question, "Is it value for money?". This we will discuss later.

The outline picture of the V7 is that it is a 757 c.c. 90-degree vee-twin with its engine mounted across the frame, the cylinders protruding upwards and sideways. It is a heavyweight of the first order, turning the scales at just over 500 lb. It does not carry its weight lightly. It looks massive. But does that detract from its looks? Not for me but, as always, this machine's beauty will be in the eye of the beholder.

The more I rode the V7 the more I grew to appreciate its looks. They seemed right for this sort of machine, with this sort of performance. It looked and behaved like a lazy giant, reluctant to lengthen its stride but content to call on its deep reserves of leg power to lope along at a deceptively high speed. It is due to no lack of enthusiasm on my part that my riding experience of big vee twins is just about nil. I did not need many miles on the Guzzi to begin to appreciate its virtues. Yes, I know I said I was going to describe the Guzzi but every time I sit down to describe the engine my mind drifts into a reverie of riding recollections and that deep, chuffing lovely power comes to mind. Back to the description . . .

The engine: Powerful not just in action but on paper with a claimed 60 b.h.p. available at 6,500 r.p.m. Sixty? That's as much as the Norton Commando gives, and much more than some very potent machines we have tried. Perhaps that 60 is pitching it a little high. The compression ratio of 9 to 1 seems almost low by today's standards. The overhead valves are push-rod operated. Bore is 83mm, stroke 70mm. The crankcase is in light alloy, of course, as are the cylinders and heads, the barrels having hard chromed bores and the heads inserted cast-iron seats. Crankshaft and con-rods are steel with AL-Tin alloy thin wall bearings. The crankshaft is bolted up, enabling big-end bearings to be replaced, provided the crankshaft does not need regrinding, without complete engine dismantling.

On the front end of the crankshaft is a helical pinion driving the half-speed camshaft gear above and the oil pump gear below. In front of this assembly is a housing for the belt drive for the generator. The belt runs from a pulley in front of the oil pump gear to the generator, which is mounted above the engine. The generator is a 300W (yes, three hundred!) Marelli unit. Marelli also make the regulator, mounted at the rear of the frame. Adjustment of the vee belt is by shimming the lower pulley, Volkswagen style. The front crankshaft main bearing is housed immediately behind the crankshaft pinion. The rear main bearing is in a detachable housing in the rear of the crankcase. The crankshaft's robustness can be seen from a few of the dimensions quoted in the workshop manual. Crankshaft pin diameter is 1.73in; diameter of the mainshaft on the flywheel side is 2.1in, on the timing side 1.5in. The crankshaft journal's inside diameter is 1.85in on the timing side, 1.25in on the flywheel side. Oil delivery is by gear pump. Operating pressure is 54/60 lb sq. in. Recovery is by gravity and the oil supply of 3¾ quarts is carried in the sump. The crankcase breathes to a breather box mounted at the rear of the engine. A diaphragm on a box comes into operation when the relief valve opens to restore required oil pressure. Behind the rear main-bearing housing the flywheel is bolted by six bolts to the rear of the crankshaft. The flywheel is recessed to contain the two-plate dry clutch. Eight springs accept the pressure and drive is by toothed pressure plates to the inside lip of the flywheel. The whole assembly is secured inside the flywheel by the starter ring gear, which is bolted to the flywheel by eight bolts. The starter motor is bolted, rather untidily, to the bottom left-hand side of the gearbox housing with an electro-magnetic ratchet control preventing damage once the engine has fired. Operation of the clutch is by a pressure rod running through from the rear of the gearbox with a caged bearing easing its path in the gearbox. The entire engine/clutch layout is not dissimilar to that fitted on the forerunner to the present-day BMW. The same could be said of the method of transmitting and driving.

The gearbox mainshaft, driven by a gear on the clutch shaft, turns the layshaft. The mainshaft is provided with four fixed gears and the layshaft with four frontal gears, two sliding sleeves and speedometer drive gear. The gear-change is by rocking pedal operated by the right foot, but it can be easily adapted for left-foot operation, with the brake pedal changing sides also. Drive from the gearbox is by a shaft which utilizes the right-hand swinging fork tube as a housing to carry it tidily to the bevel box. At the gearbox end the cardan shaft is connected to the gearbox via a universal coupling housed in a rubber gaiter. The shaft is a comparatively short unit, splined at either end. The front end is a secure push fit into the universal coupling and the rear spline fits into a tubular splined

Transverse vee twin is said to produce 60 horsepower, on a compression ratio of 9 to 1. Clutch and gearbox layout is not dissimilar to that of earlier BMWs, clunky change is rather like, too

Italian liking for warning lights in varied colours is displayed on the Guzzi control panel. All castings and machining in this area are of a very high standard

housing which also accepts the spline on the rear bevel pinion. The bevels drive on to a bevel-cut crown wheel giving, in the case of the test machine, a final drive ratio of 4.375 : 1 (8/35). A set of lower sidecar ratios is available in Italy but the test machine relied on standard solo ratios. The rear-drive housing is located on the right side of the machine, with the brake apart from the unit on the left side. The concept of the Moto-Guzzi engine unit is one of solidity and, from the look of it, long life. The engine looks heavy, robust and reliable. In our earlier comments about prospective owners' concern over spares, we should have mentioned that many parts of this machine are Fiat, the starter motor and dynamo for example. This could well affect availability and price considerably.

One last look at the engine before we pass to the rest of the machine. Carburettors are 29mm Del 'Orto. The distributor protrudes from the rear right-hand side of the crankcase, being driven by the camshaft via a skew gear. Inside the distributor cap is the automatic advance and retard mechanism, a pair of bob weights centrifugally actuated. Ignition timing and correct functioning of the a. and r. mechanism can be checked by a stroposcope illuminating marks on the generator pulley in relation to the rear of the front timing gear case. Tickover speed is at the fixed advance position of 10 degrees to t.d.c.; at 2,200 r.p.m. an intermediate mark on the pulley should be illuminated by the stroposcope indicating that the engine has reached 20 degrees advance to t.d.c., and the full 38 degrees maximum advance will be indicated at 3,600 r.p.m.

From the top of the distributor the high-tension lead runs to the coil with the condenser affixed to the side. Altogether the engine owes much to car practice, a not unreasonable situation as this engine is said to have started life as the power unit for a small Italian army cross-country truck.

Turning our attention to various other parts, it is not the frame which contributes most to the bulk of the V7, but the front mudguard. It is of considerable size and substance, supported by robust stays and shrouding a front tyre which, at 4.00 × 18in, is one of the largest we have had on a test machine. Such a tyre might almost have been designed exclusively for use with a sidecar. The frame is of twin-cradle type, substantial, strong and clearly well up to the task of coping with sidecar stresses, as the makers acknowledge by the provision of sidecar lugs on the sidecar version (also available in Italy).

The petrol tank is one that is in keeping with the rest of the machine. The popular two-gallon tank favoured on the other side of the Atlantic would look ridiculous on the Guzzi and the makers have sensibly opted for a good-looking receptacle that holds just under five gallons with, so "the book" declares, a gallon in reserve. In this respect our tank would not play ball and refused to yield another drop of petrol once reserve was switched on, although there was clearly a fair amount left in the tank. Not for the first time this year I had to find this out the hard way. Luckily my wife was in the sidecar and she was able to push while I steered. Come to think of it, this may have jaundiced her view of the whole outfit.

The most noticeable omission from a machine which is intended for at least occasional sidecar use is a steering damper. More of that later. The dualseat is firm, in traditional Italian style, but was as comfortable as one could wish, this happy result owing something no doubt to reasonably soft suspension and fat tyres. Two grab handles are provided for a pillion passenger. Playing its part in the overall weight of the V7 is a 12-volt battery, which at least ought to make for easy replacement, for it is of car type and would be, one imagines, available in any battery depot at a price much less than that which motorcyclists are normally expected to pay. It runs the full width of the machine and is tucked tidily behind two removable covers, behind which also lies the air filter, a substantial paper-element type. Out of character are the exposed rear springs

Front brake, beautifully made and visually very impressive, was hard put to cope with the outfit's near-1,000 lb total laden weight

on the suspension units. They are neither in tune with the touring concept of the machine nor do they blend with the robust front forks.

It is fashionable these days to fit a balance pipe in the exhaust system but Guzzi have changed the formula a little and built the pipe into the silencers (£28 the pair if you have to replace them), running it across just before the silencer joins the exhaust pipe. Front brake is two-leading-shoe, rear a conventional single ls unit.

The handlebars might not have pleased us on a solo but for a sidecar outfit they were ideal. The controls were very light. All electrical controls are operated by the left thumb, with a master key in the centre console to activate the ignition or the parking lights, with the key being removable then. The starter motor button is in the right-hand throttle assembly. It was a little on the sticky side and did not always make contact at the first attempt. Also on the right side is the air lever which, by British custom, worked the "wrong way", being open on a slack cable, closed on a tight one. Easy to get wrong first thing in the morning until you get used to it.

Very partial to their little lights are Italians. Sit in any Italian car and an array of lights greets you—blue ones, red ones, green ones. The Guzzi is quite mild by those standards, I suppose. It has four. Two red ones, ignition and oil-pressure warning, a green one for the lights and a yellow one for the neutral indicator. They missed out on the main-beam warning light somewhere. A somewhat bulky centre console completes the handlebar layout, with speedometer (calibrated in k.p.h. on our model) and tachometer between it and the headlight. The console has the slight disadvantage of making the fitting of a steering damper difficult.

Finished in white with red and black lining, and with a black frame, the Moto-Guzzi is not the sort of machine that one could pass by and not notice. Black and white may not seem very attractive but, with the enormous engine assembly finished in matt alloy and the rear drive in polished alloy the finished product is far from an ugly duckling. On the contrary, it is a very handsome machine.

One occasionally sees a wedding where a beautiful bride glides up the aisle on the arm of a groom who, at best, can be described as "interesting" looking. The marriage of the Astralite to the Guzzi might well be described in this vein. Even its builder would not claim it is the best-looking sidecar outfit on the road. To be fair, that was not the object of the exercise, which was to show that the Moto-Guzzi is as good a sidecar machine as there is on the market. Choice of sidecar was dictated by availability, as I've indicated. There was an Astralite in the workshop at Barretts. Had there been a Steib or a Palma we might have been singing the praises of the best sidecar outfit around today. As it is, we are stroking our chins and saying "Mmmm, very interesting". The sidecar is fitted to the V7 at four points. The front swan neck utilizes the crash-bar fitting very conveniently. Below that, the other front fitting goes to an alloy block fitted between the engine and frame replacing, we imagine, the spacer on the front engine bolt. At the rear both fittings are clamped to the frame, one alongside the toolbox, invading the space needed by the cover, the other to the bottom of the frame by two clamps. We had no complaints whatsoever with all this; the outfit handled like a sports one. Our only comment is that proper sidecar fittings would make a very much neater job. Presumably the sidecar version frame would overcome most of these problems but it does not seem likely that the sidecar lugs would be on the left, bearing in mind how few countries use sidecars on this side. A brake was fitted to the sidecar but was not connected up for use.

Forgive us dwelling so long on the bike but we do not often get our hands on a Moto-Guzzi, so we lingered over each feature with the interest of a child getting a new toy for Christmas. It is certainly among the top five motorcycles in the world. Which makes it all the more surprising that there are so few of them around. It was amusing to watch the reactions of fellow-riders on the road: the sheer amazement when we stopped at lights, say, and they first took in the V7, then let their eyes drift, incredulously, to the sidecar. Considerable

Part of the scaffolding that's necessary to ensure a stable relationship between sidecar and bike. Unsightly, certainly, but in this case extremely efficient

numbers of "civilians" were unable to resist a chance to talk about it. In many cases, although they were all once motorcyclists, the V7 was a new one in their book. Harley-Davidson and Vincent were their usual first guesses. (Sorry about that, Vincent owners, especially after that lovely Dinner you provided a few weeks back!)

How quickly one forgets how to handle a large sidecar outfit; and how soon one learns again! The biggest difficulty as I pulled away on the Guzzi was in seeing what the sidecar wheel was doing. From the pilot's seat it just was not visible and at first I was tending to give anything on that side the widest possible berth. A greater trial was to come. The road away from Redhill is somewhat bumpy and, as the steering gave a little flutter, I reached forward to give another turn on the steering damper. No steering damper. The flutter at low speeds rarely developed into anything worse. Rarely but . . . once or twice, if one relaxed at the moment the power was turned on after changing gear, the flutter became a threat of something more. The answer seemed to be to keep the power turned on but theory is all very well when you've just frightened yourself to death. Final solution was to keep a very firm grip on the bars at such times!

It took a few miles to convince myself that I was not in charge of a monster, and from then on I was able to relax and enjoy myself. I did not expect the solo gearing to give any real problems for I meet the same problem without worries on my own machine occasionally. Barretts had retarded the ignition a little to ease the burden, and it pulled away like a steam engine, the power of the vee twin coming in right from the bottom of the rev range. If I desired, the engine would chugg away like a steam engine. Not exactly what the designer of the V7 had in mind, perhaps, but useful nonetheless.

Someone had done a lot of work getting the alignment of the outfit right and it pulled away in a straight line with little corrective effort needed by the driver. Changing gear was exactly like that on any other engine-speed clutch machine that I have tried—"clunky" if one was less than dedicated. Even with absolute concentration the downward changes, especially into second, were liable to clash. I tend to blame this on the rocking-pedal gear-change for it prevented the rider using necessary delicacy and feeling the gear in. With toe or the heel it was never quite right. (There may be a deal to be said for having the optional gear-change side in operation for some riders are quite unable to adapt to the opposite side, and many customers must be lost this way.)

It is, perhaps, a little unfair to draw too many conclusions from the performance of the brakes. When we tried the model, briefly, in solo trim a year or two back, there was no obvious lack of power, but the addition of about 200 lb of sidecar had its effect on them and they were hard put to cope with anything really demanding. And the sidecar brake was not connected.

Barrett's Guzzi/Astralite is really a mobile test bed; there have been no concessions for sidecar use. The suspension was still on solo springs and, as we have mentioned, the gearing was solo. This showed itself, as experience had told me it would, in top gear at speed. Even in solo trim the Guzzi was overgeared; you may imagine, therefore, what a sidecar did for it. Top speed in top gear was occasionally as low at 55 m.p.h. and it took reasonable conditions to have the speedometer needle up to 100 k.p.h. (62 m.p.h.). A fairer assessment of the machine's potential may be obtained from its performance in third gear. In this gear the V7 would buzz happily up to 125 k.p.h. (77 m.p.h.) and stay there, the tachometer indicating 5,500 r.p.m. A good thing about the high top gear was that the outfit would chuff along on a whiff of throttle at 60 m.p.h., showing 3,800 r.p.m. But the slightest gradient would call for third. At the other end, top gear could be held down to

Continued on page 131

MOTO-GUZZI V7 SPORTS

THE MOST BEAUTIFUL TEST-BENCH IN THE WORLD

BY JAN LEEK
PHOTOGRAPHS: LARS BERG

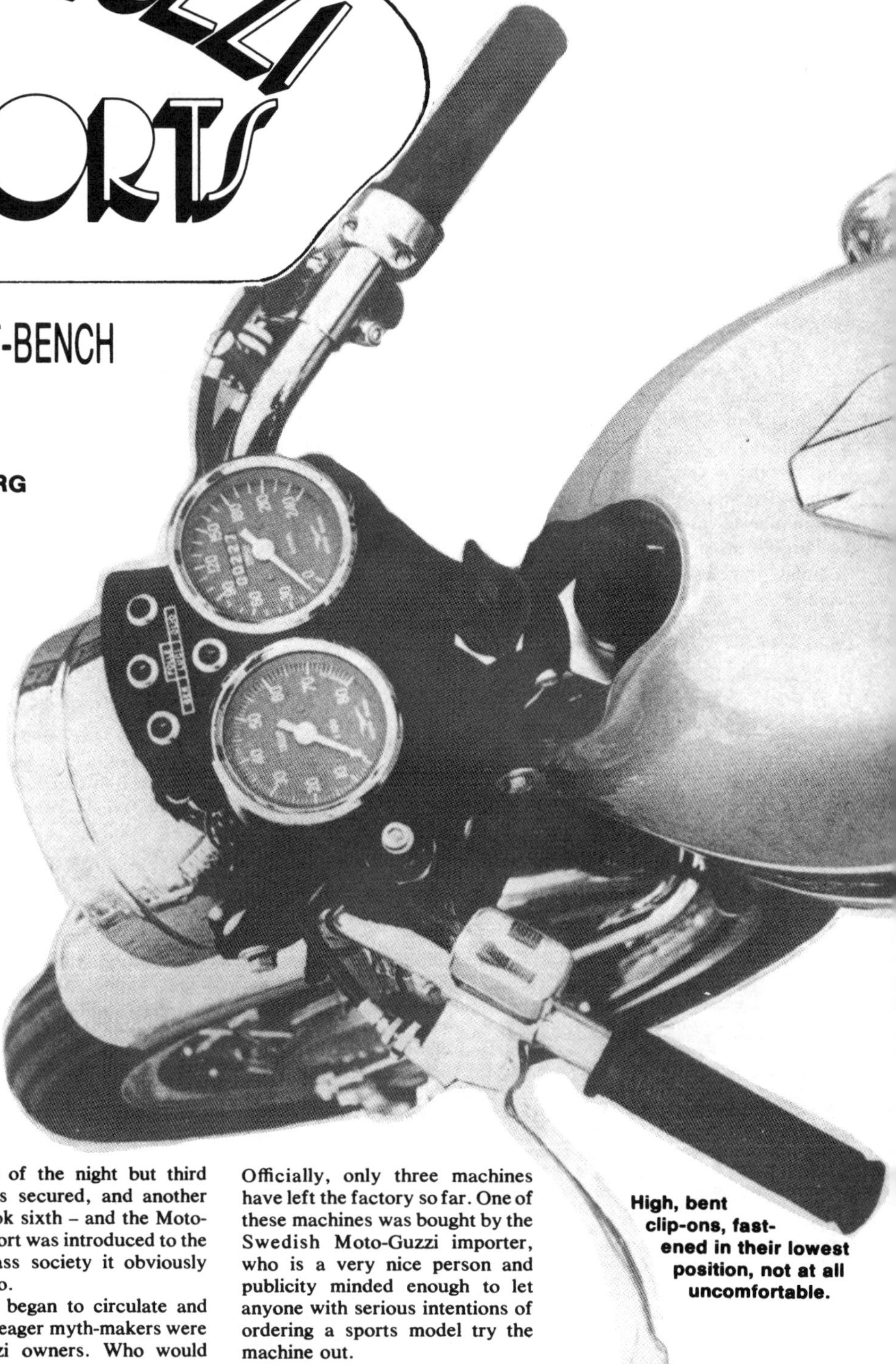

High, bent clip-ons, fastened in their lowest position, not at all uncomfortable.

THE TRADITIONAL CONCEPT of an Italian motorcycle is of something red, low, light and awfully fast. Temperamental. Something that speaks to your senses instead of your rational intellect.

Therefore it was that much more astonishing when Moto-Guzzi released their V7 a few years ago – a sturdy machine obviously built for long distance touring, and only for that. Of course, in Italy lots and lots of motorcycles are made for the only purpose of transportation, but they are seldom exported and this is where the new Guzzi differed from its brothers. It was exported, and in a very big was. Even in the central European countries, where they are so consciously aware of mechanical finesse, the machine was a big hit.

The years passed, and the reputation of the V7 had its scars, but it survived, with its basic reputation as a long distance "bus" intact.

Production machine racing seemed to be out of this machine's reach, but in 1970 the factory entered two 'bikes in the 24-hour marathon at Le Mans. The machine is heavy and not a happy revver, but finished ninth, which must be considered a success whichever way you look at it. But, then the rumours started. . . .

A new, lighter machine is on its way. The people who knew nodded secretly: "Just you wait!"

No official statements were made and the public was still ignorant of the machine's existence when it entered this year's 24-hour race. A broken rocker arm and a puncture prevented the machine from winning after leading most of the night but third place was secured, and another Guzzi took sixth – and the Moto-Guzzi Sport was introduced to the upper class society it obviously belongs to.

Photos began to circulate and the most eager myth-makers were the Guzzi owners. Who would have thought that *they* were the ones to ride around with a potential racing machine?

Production is not due to start until the beginning of '72 and the first deliveries to take place in April. Until now, only a small number of machines has left the factory and they are intended entirely for PR purposes. Officially, only three machines have left the factory so far. One of these machines was bought by the Swedish Moto-Guzzi importer, who is a very nice person and publicity minded enough to let anyone with serious intentions of ordering a sports model try the machine out.

Scandinavia in the middle of October is usually damp and cold. Last year the calendar had missed something as the temperature was awfully close to freezing point and the sun was only there to give an impression of ice-blue sky, to go with the temperature.

Rumours had it that the frame was red and the tank green. It seemed doubtful if it was worthwhile wasting much colour film on something which looked like it came from the candy store. Nervous! But when we got there, outside the workshop there was this candy bar. Red and green. And if I had had the money at the time I would have spent it all on red and

Overall impression is of a fast, serious machine. Fast enough to interest those only concerned with production racing.

The front brake is new, with twin leading shoes on each side.

green without hesitating a moment. *There* was an Italian motorcycle! The sports version is low; with the same wheelbase as its big brother it also seems long. If the V7 is styled after the '50s idea of a well-shaped object, Marilyn Monroe-like, the "Sport" is a product of the Seventies; hard, straight, a machine with a message. The frame is a completely new construction and the rear part shows its racing inheritance. The front forks too – polished aluminium, open Ceriani type, but made by the Guzzi factory. The front brake is also new (even the fastest guys like to stop now and then), double-leading shoes on each side. As a rear brake the old double-leading front shield is put into the old rear drum.

The narrower tyres (3.25/18, 3.50/18) in conjunction with the new, light, chromed mudguards, make the 'bike look lighter as well as making it steer lighter. The handlebars are of a new type. High, bent clip-on bars, fastened in their lowest position give a semi-crouch, not at all uncomfortable and rather well suited for fast road work.

Also new is a hydraulic steering damper, not at all uncommon on sports machines today, but they're so nice to have, it's worth noting.

The ignition lock has a somewhat strange place on the 'bike. It sits on the top frame tube, where the tank is divided up front, and is combined with a steering lock. When you pull out the key and turn the front wheel all the way to the right, you have also locked the machine. And it also means you have only one key to lose.

The tank is Italian. Period.

The sub-frame can be split from the rest of the frame and the engine simply dropped down and let free for whatever work is necessary.

At the rear, Koni dampers spring the wheel and are a nice example of the factory's willingness to improve their own product, even if it means that they have to call for help from outside.

The footrests are mounted in a long piece of moulded aluminium, bolted to the rear frame, and are in good proportion to the relatively low handlebars.

The seat is hinged at the rear and below it is hidden a battery of the same enormous size as the one that takes up so much space on the standard V7. Together with the battery are all those fuses which keep the electrical powerhouse under control. Now for a nice point. If you turn on the parking light and lift the seat, a contact will turn on a small lamp so that you can see to do whatever you feel like doing down there. Thoughtful.

The new silencers give a nice, low purr, not too high, not too low. Somewhere between a standard Commando and a BMW. Just right.

Between the pipes there used to be a balance pipe. Unfortunately high revs could create vacuum and the intended effect never resulted. To correct this, two pipes, crossing each other, are now just in front of the silencers.

The gearbox and engine castings are mostly new. The link mechanism to the gearbox is a much better-working contraption and it seems as if a few things have happened *in* the box as well.

The most obvious modification is the alternator mounted directly on the crankshaft and moulded into the casings like a Honda unit. 180W gives light and juice to the battery.

Compression ratio is 9.8 : 1 and a new camshaft helps mix the sauce to give the nicest dish you can think of. Although it is not official, it is obvious that a lighter crankshaft and lighter valve

MOTO-GUZZI V7 SPORTS

mechanism are only some of the more apparent modifications made.

Carburetters are the same as before, but one millimetre bigger to give a choke size of 30mm. Unfortunately no air filter elements were present in the rubber box where everybody expected them to be, and a gasping sound could be heard from the engine at low revs. This is not the most significant objection to be raised, as the hardchromed bores in Italian products have shown themselves to be of a quality less than first-class. With no air filters, well. . . .

Rattles set in

FOR NEXT YEAR A regulator for the starter engine is standard on all Moto-Guzzi's models and will probably give the battery a long and healthy life. Alas, 'twas not always so, especially in colder climates. This one has it already and the engine starts quickly and sets down to a quiet tickover. No noticeable side movement, just a sound. All new Guzzis are usually quiet in the valve mechanism but after a few hundred miles a rattle begins that follows the machine all through life. This one was quiet when I rode it.

The valve mechanism has to be lighter on a machine with the red line at 7200 rpm, and it is possible that the Guzzi rattle has gone for ever.

The brake pedal and gearchange mechanisms follow the English pattern and will probably have to be changed if it is going to be exported. That means first up – no clunk – try again! Release the clutch and, yes, it moves. Somebody spent overtime working on a gearbox last summer.

Whip open the throttle and, with no gyroscope effect whatsoever, the machine moves forward, no tendency of tilting to the left, not even when second gear is pushed in and more throttle raises the revs to 5-6-7000. Straight line? Yessir!

First gear is rather high and 30 mph is a comfortable speed at about 4500 rpm. Second is so close that the first time you think you have missed a change, but then you feel the engine pulling strong and mightily to another gear change.

Without access to proper testing equipment and a track (without speed traps), all our riding had to be done on public roads, although we chose the straightest we could find.

At lower speeds in first and second gears, 3500 was a reasonable lower limit, but already in third gear 4000 proved necessary and in fourth the limit was raised to 4500, which goes to show that this is not the low torquer one believes a driveshaft, V-twin engine to be. Yes, there is a fifth gear as well. But we'll come to that.

My Porsche-driving, photographing friend was leading the way out of town and looking anxiously through his rear-view mirror to take in the sight of whatever he thought was going to happen.

Open roads, sunshine, feel the engine and let it accelerate the way it wants to. Fourth gear – 4000, raise to 5000 and get fifth. Nice! Awful wind blowing, though. Suddenly, a little rational pixie behind my left ear told me something. 5000 is a lot of revs on any machine and this is a seven-fifty!

Oh, my . . . that speedo is reading about 95 mph, and this is a unique machine, and the money in your wallet, and. . . . And I slowed down.

It suddenly dawned on me that the 'bike had been doing an awfully good job in those bends back there. To find out what the machine was capable of doing, I first tried out a medium range. 5500 rpm seemed a nice place to change gear without stressing the engine, so the most workable range for the engine proved to be 4500 to 5500. In that case 5000 is a good medium range for comparisons between the gears (I did not want to flog the machine too hard as it had only just been delivered from the factory, although it felt loose enough to permit a short burst of revs). 5000 in first showed to be about 35 mph (all readings were done after the instruments on the machine and can only be approximate), in second, slightly more than 50, and in third, close to 70. Third gear was very useful and could be used from 45 to 70 mph as it gave a good throttle response and did not sound stressed or felt like it was held back or wanted more stress. Just so . . . civilised.

Fourth gear felt like fourth gear on the Ambassador and 4500 or 5000, where it felt free and happy, proved to be about 80. As you probably know by now, that fifth gear is high. Below 75 there is no use for it. Belief in a basically good design and an engine that felt happy made me give the machine a short run to prove itself. A long straight was found and after fourth gear, 90 mph and roughly 6000, fifth gear was engaged.

From a little more than 5000 (difference between fourth and fifth is about 800 rpm) how that engine pulled, right up towards 7000. With the needle swinging towards a certain 125 mph and the rev-counter at 7000 I was sure that

***There's* an Italian motorcycle – hard, straight, a machine with a message.**

Moto-Guzzi muscles. Is the factory simply "out on the beach" to show them off?

the factory's claim that every machine leaving the factory will be guaranteed for at least 128 mph (206 kph) is true. Remember, this was without a fairing and the machine was as delivered, right out of the crate.

At no time was it possible to experience a feeling of speed. The 'bike handled beautifully at all times and the only way to know how much fun you had was to read the speedo and tell yourself: "This is going fast!" The machine feels long, like the Ambassador, but gave the impression of having a quicker and more direct steering. If this is due to the smaller tyres or if fork rake is different, I don't know, as there were no facilities available for measuring the fork angle or rake.

The overall impression is of a fast, serious machine. So fast, in fact, that it is assumed that it is of interest only to those who wish to compete in production machine racing. Another viewpoint is that the factory is just out on the beach to flex its muscles and show that it can still do it. If this is a product of the PR department only, then it is an expensive way of proving itself. With a small production output planned (300 units have been mentioned), this does not seem far-fetched but it seems improbable that, with a good market response, production should cease entirely.

Details show that the road rider has been thought of and it is safe to suspect that this machine will be built in larger quantities. If not, it is probably the most expensive image a factory has ever created, and the most unnecessary.

Some parts on the machine may very well be included on the V7 Ambassador, and if so, the Sports version is the most beautiful test bench in the world.

Front forks in polished aluminium, Ceriani type, are made by the Guzzi factory.

SPECIFICATION

ENGINE: Bore and stroke, 82.5 x 70.2mm. Capacity 748cc. Power (claimed) 70 bhp at 7000 rpm.
ELECTRICS: 12-volt 32 a.h. battery. Twin coil ignition, double points and condensers.
WEIGHT (dry) 453lb (206kg).
WHEELBASE: 59in (1470mm).
WIDTH: 28in (700mm).
GROUND CLEARANCE: 11in (280mm).
SEAT HEIGHT: 31in (770mm).
WHEELS: Front, 3.25 x 18in. Rear, 3.50 x 18in.

MOTO-GUZZI V7 AND SIDECAR

Continued from page 127

25 m.p.h. with no objection from engine or transmission. Transmission: by shaft drive. Civilized, clean, smooth, reliable. You get the impression that I like it? You're right.

The test machine was not the usual (?) gleaming, brand-new, just-out-of-the-showroom model spoiled road testers tend to expect. It was a second-hand model from stock, with 13,000 miles on the clock. It had been well used. The engine was not as quiet as it might have been—valve-gear noise mainly—and there was considerable clutch chatter. The exhaust note was no more than a whisper—so well muted, in fact, that one had the feeling that the tiny apertures at the end of the silencers were barely adequate. But these silencers conform to the stringent silencing laws in Italy, and a less effective pattern is available for the United States, we were told. But the standard silencers enabled the outfit to be ridden quite exuberantly without giving offence, and it handled so well that after a while it was possible to treat the Guzzi as though it had a Steib on the side! The longer one drove it the more fun it became, with just this slight niggle of not being able to utilize top gear to best effect. The riding position was exactly to my taste. Large, comfortable handlebars, firm but comfortable dualseat and suspension that, although on solo springs, did not wallow but absorbed all road irregularities with ease. Night driving was reasonable, although the lights were not quite as good as we would have expected from a 12-volt system using a 45/40 bulb. The horn was fair. No more.

In spite of the attempts of the petrol tank to defeat us we calculated that the outfit was returning 35/40 to the gallon, depending upon use. We have had worse from solos.

There was a great deal of pleasure in driving the Moto-Guzzi, even allowing for the too-high gearing. It could be driven through corners very quickly, with apparently no chance of the sidecar wheel lifting. Handling, in fact, was very good indeed.

The Moto-Guzzi V7 is an excellent motorcycle in its own right. And, as our few weeks with the outfit showed us, it is one of the most suitable bikes around today for serious sidecar use. The spread of power and low torque of the 750 vee-twin are ideal for the job. The combination of engine/exhaust note was not unlike that of a Harley Davidson, a deep, whoofling burble. It started with extreme ease, the engine pulled the 750 lb+, and the two-plate dry clutch coped manfully. The engine unit remained completely oil tight during the course of the test save for a little oil on the sump around the hole under the clutch housing. A leaking rear-main oil seal perhaps? The generous front mudguard was every bit as efficient as it looked and kept the front half of the machine admirably clean, even after use in the wet. All the controls were light and easy—with the possible exception of the rocking gear-change pedal. A very pleasant, powerful machine that we liked very much, but—now comes the final question. Is it, at £950 (the concessionaires are hoping to reduce this to £900 pending negotiations with the Customs), value for money?

Our biggest, in fact our only real, criticism of the Guzzi is that it is surely a little "over-constructed"? It should be possible to save weight (and, consequently, cost?) by a little careful streamlining. Items like the starter motor tacked on the outside, a sort of afterthought, give the machine a slightly "agricultural" appearance. Perhaps this is due to its ancestry as a power unit? To return to our earlier question, we have come to the conclusion that, if it sold for, say, £700/750, the Guzzi would be seen much more frequently on our roads. As it is, its appeal will be to the man who is prepared to pay that little bit extra for a machine of character. For, make no mistake, the Moto-Guzzi is a machine of great character and not a little charm.

B. P.

The large-capacity motorcycle most talked about during the last few months in Europe is the Moto Guzzi V7 Sport. Derived from the well known Ambassador, this model has been specially designed for sporting enthusiasts and is perfectly suited for the speed and endurance races which are becoming so popular everywhere.

Engine design is the same as the more familiar touring and Police versions with the cylinders placed transversely, making a V of 90°. One would think that this kind of engine configuration is not very adaptable to a strictly sporty machine, but an entirely new frame design eliminates any thought of it being an only slightly modified tourer. Consisting of a solid trellis-like construction with the lower cradle members detachable for easy engine installation, it resembles the MV Agusta of Grand Prix fame.

Compared with the Ambassador, the Sport model differs in many respects. The alternator is placed on the front of the crankcase which completely revises the design of this part of the engine. Other differences include finning, engine block casting and the shape of the oil sump. Of interest to competition-minded riders is the fact that the capacity has been reduced from 757 to 748cc, which will allow it to be raced in the AMA 750cc class as soon as the required number of new machines are produced. New design cam, 30mm carburetors incorporating an accelerator pump and a twin-breaker distributor result in an output of over 70 hp at 7000 rpm. Fitted with an electric starter, the total dry weight is 454 pounds. A five-speed gearbox and heavy-duty clutch allow a maximum speed of 128 mph.

Giving a fast look at the machine you'll note not only its compactness, but the exceptional ground clearance desirable for high-speed riding. The handlebars are clip-on type, adjustable for either racing or touring. Another interesting feature is an anti-theft device connected to the hydraulic steering damper. The ignition key is located in the opening in front of the fuel tank, while the electric starter button is on the right-hand side just beneath the throttle. Removing the key locks the handlebars to either the left or the right. On top of the front forks are warning lights, speedometer and tachometer. Looking carefully at the machine you can also see that the rear fender is hinged for easy removal of the wheel. The battery and fuse box are located under the flip-up seat which has its own unique courtesy lamp.

The front fork is the usual telescopic type, but it is of Moto Guzzi design and manufacture. The completely new front brake, also made by Moto Guzzi, is a 230mm unit incorporating four leading shoes, while the rear brake is the older-style twin leading shoe as used on touring models. Unusual items are the gear lever and the foot brake pedals. The brake pedal is rather long in order to allow a minor foot pressure

MOTO-GUZZI'S NEW 750 SPORT

3cc's smaller and 2 seconds faster.

Report and photos by Roberto Patrignani

1

1. The Sport model doesn't look much like the Moto Guzzis we're used to looking at. The engine is derived from the more familiar model but all else has undergone extensive change in interest of speed. *2. Moto Guzzi engineer, Lino Tonti, shows off the latest creation to ex-world Champion road racer Bill Lomas. In the Fifties, when Moto Guzzi was actively involved in racing, Lomas rode famed water-cooled eight-cylinder models.* *3. Perhaps more familiar is the flashy Mike Hailwood. He test-rode the 750 Sport and was most impressed with both the performance and handling.* *4. Low profile results from all-new frame and forks allowing lower placement of engine unit. Finish is red frame and yellow fuel tank.* *5. Right-side view reveals massive brake hubs and the shaft drive through the swinging arm to rear wheel. Note adjustable clip-on bars that can be raised, lowered.*

2

3

4

5

when braking. Footpegs are the folding type. By manufacturing all these parts themselves, Moto Guzzi demonstrates the care with which the V7 Sport has been built.

When riding the V7 Sport you can feel the immediate and complete security of a machine which follows the pilot in each movement even when pushing it at high speed on curvy roads. Mike Hailwood rode it recently at the Monza racetrack, and as several English newspapers reported, he was very enthusiastic about it. This model has already started to be sold in Italy and even if it costs about $2500 it is very much in demand.

All this enthusiasm for a motorcycle derives from a fundamental fact which you can feel while riding it. It has exceptionally good handling and stability because it is so compact and has such a low layout. The frame is a real sporting type and its designer, Lino Tonti, is well known for his art in this specific field.

The V7 Sport is also easily braked, its gearbox is efficient and precise, and engine vibration can only be felt when idling. Shaft drive is a luxury to which you must become accustomed, for when shifting down through the gears things seem less elastic than with chain drive. Acceleration is very good even though the gear ratios are widely spaced.

Although the V7 Sport has sporting features it can be used for touring and long trips without tiring the rider. This is due not only to the aforementioned characteristics but to some less noticeable features such as an engine that does not wobble, plus sturdy suspension. The V7 Sport is really a motorcycle which means something in today's motorcycle world, and even if the price is a little high it will surely be accepted by two-wheeled enthusiasts everywhere. ●

MOTO-GUZZI 750 SPORT

1

2

1. Though similar in external appearance, the engine is really quite a bit different from previous models. The alternator is mounted on front of crankshaft and the sump and crankcase castings have been slightly changed also.
2. Extensive tests at both the factory and in local Italian production machine races have proven the V7 Sport to be both fast and dependable. This shot is at Monza 'Drome.
3. After a look at the insides it's not surprising that the engine is trouble-free. The one-piece crankshaft is forged steel and accommodates both rods on a common journal. Camshaft gears are plenty hefty too. Cam is new grind aimed at all-around performance for street and track.
4. Most interesting change is the reduction in engine's displacement to get it under 750cc. Old style was a hair too big to qualify for 750 class. Aluminum cylinders use chrome bore for long wear and better heat transfer. There's no liner at all, just chrome on the aluminum. Lightweight pistons use four rings, ride on large piston pins.
5. Though the cylinders are interchangeable, heads are not. Also cast in alloy, they are light and have inserted bronze valve seats. Valves are pushrod-operated as in street machines but extra attention is paid to chamber and ports.
6. Drive to the five-speed transmission is through a twin-plate, heavy-duty clutch. Five speeds are new for '72, could become option for the street touring machine.

3

4

5

6

MOTO GUZZI V7 SPORT 750cc TECHNICAL SPECIFICATIONS

ENGINE: Four-stroke 90° twin.

CYLINDERS: Light alloy with hard chromed barrels.

CYLINDER HEADS: Light alloy with inserted valve seats.

CRANKCASE: Light alloy.

BORE AND STROKE: 82.5 x 70.2mm.

CUBIC CAPACITY: 748.388cc.

VALVE ACTUATION: Pushrod and rockers with central camshaft.

CRANKSHAFT: One-piece forged steel unit with the conrods attached to the same crankpin.

HORSEPOWER: 70 hp SAE at 7000 rpm.

COMPRESSION RATIO: 9.8:1.

IGNITION: Twin contact breakers, twin condensers, and twin H.T. coils. Marelli 275 spark plug, or equivalent.

STARTING SYSTEM: Electric.

LUBRICATION: Pressure by gear-driven pump.

RECOMMENDED FUEL: High octane gasoline.

FUEL TANK CAPACITY: 5 gals. including 5 pints reserve.

CARBURETORS: 30mm choke twin carburetors with concentric float bowl, accelerating pump, and inlet silencer.

CLUTCH: Twin dry discs, on flywheel.

GEARBOX: Five speeds, foot-operated on the right-hand side of the machine.

TRANSMISSION: Primary drive by helical gears, final drive by constant speed double joint shaft and bevel gear set.

FRAME: Duplex cradle.

SUSPENSION: Telehydraulic front fork, rear swinging arm, with adjustable telehydraulic shock absorbers.

WHEELS: Light alloy rims, 3.25x 18 front and 3.50x18 rear tires.

BRAKES: 220mm four-shoe front brake, twin leading shoes rear brake.

ELECTRICAL EQUIPMENT: 12-V battery with alternator. Fuse box under the tiltable seat. Courtesy lamp. The system is wired for the installation of turn light indicators.

OVERALL DIMENSIONS AND WEIGHT: Wheelbase, 57.87"; length, 85.23"; width, 27.55"; height to panel, 40.75"; ground clearance, 11.02"; seat height, 30.31"; dry weight, 454 lbs.

PERFORMANCE: Fuel consumption 30 mpg. Top speed 128 mph. Standard ¼-mile in 13.12 secs.

MISC.: The V-7 is fitted with a two-piece handlebar adjustable for height and width, instrument panel incorporating speedometer and tachometer, steering lock key, and folding footpegs.

CYCLE ROAD TEST:

MOTO GUZZI: V850 ELDORADO V750 SPORT

Among motorcyclists, there is a small group that enjoys riding on long trips. It isn't a group, actually, for you seldom see more than a pair who have covered more than the same thousand miles on a trip at the same time. And more often than not, the long-distance rider is a loner. Alone at times by choice, and in other instances simply because it's awful hard to get any combination of two bikes and riders together that are comfortable with the same pace.

Some of these cycling tourists like to break up their trip with social and/or service stops at dealers' shops along the way. Others don't particularly cotton to the hail-fellow-well-met joviality at the dealer's, but are compelled to endure it by their bike's lack of reliability or frequent maintenance schedule. And another kind of rider thoroughly enjoys working on his machine: he totes a full tool kit with all the special oils and greases that his bike continually distributes on him, itself, and the road.

There are lots of motorcycles which provide adequate service to the above types. Today, it is rare to find a new big bike that sheds parts steadily along the road, and not too many vibrate so badly that the rider has to rest every hundred miles. A bike as small as a 450 Honda is an adequate tourer for a solo rider who doesn't carry much of a pack.

But for the cycling tourist who wants to be free, there are only two motorcycles to choose from: the big Moto Guzzi V-twin, and the BMW R75/5. Freedom means a lot in this sense. Freedom for a big rider to carry a big passenger and the necessary luggage, even in a side-car if desired. Freedom from the chain which the rider had to stop every 200 miles to lube or adjust or both. And freedom from having to plan a journey route around the business hours and location of dealers which have the special tools or stock the spare parts and special oil that some bikes require. In all, freedom to simply ride wherever and whenever you wish and be able to feel good about the trip.

* * *

We've tested two models of the big Moto Guzzi 750cc Ambassador: in the Oct. '69 and Apr. '71 issues of *Cycle.* We knew the character traits, strong points, and drawbacks of these majestic tourers. In the first part of this test, we'll refresh your memory of them and tell you what we think of this year's improvements. In the second part, we look at one of the most unlikely superbikes imaginable, the V750 Sport.

PHOTOGRAPHY: JIM MCGUIRE

At first glance, the V850 Eldorado looks exactly like last year's V750 Ambassador, and not very much different from the 1970 model. The improvements, then, are all evolutionary ones that make an already great machine even better.

The Eldorado is a massive looking motorcycle, but the big fenders and wide, transversely mounted V-twin engine are deceptive. Its 58-inch wheelbase is a half-inch shorter than the Honda 750 Four and H-D Sportster, and identical to the Norton Commando. An H-D 74 is 3.5-inches longer and a BMW R-75/5 is 3.5-inches shorter.

Sitting on the best seat in motorcycling gives you added visual impression that the bike is bigger than it is. The fat, gracefully bulbous fuel tank is 11 inches across where your knees touch the chrome plated side panels, and it flares out to a width of 14 inches at the widest section. The Guzzi weighs 559 pounds with all its lubricants and a half tank of gas—almost 80 pounds more than a Honda 750. You can feel some extra heft when you let the tank fall from side to side between your legs, but the impression isn't overwhelming. The V850's heavy parts, the crankshaft and gearbox, are carried very low in the frame, which makes the center of gravity correspondingly low. The center stand lifts the bike a full two inches at the rear wheel, and we thought that it would be very difficult to put the stand down. But once you get the knack of stepping on the stand's end while tugging on the seat rail, it's actually a fairly easy, if awkward, operation. Foot tabs allow the centerstand to be operated from either side of the bike.

Our test machine was finished in solid black, with subtle white pin striping on the tank, fenders, side panels and tool boxes. The bike is also available with a white, red, or black tank with white fenders. We liked the solid black best—it gave the bike a Rolls Royce look of bulky elegance. The other schemes seemed just gaudy enough to detract from the stateliness, though the all-white wasn't bad. Guzzi is very subtle with the use of chrome and polished aluminum on the Eldorado. The rockerarm covers and generator-drive belt cover are polished alloy, while the exhaust system, cylinder guard rail, brake and shift levers, seat grab rail, fuel tank knee pads, shock springs, and front fender guard are the main chrome plated parts. All of the bolts and nuts, as well as the oil filler and inspection plugs, are chrome or cadmium plated. On one of the first nights we had the 850, it was left outside. Of course it rained like hell. The Guzzi is the first motorcycle we've tested that didn't have little rusty places all over it after a good drenching.

For some reason, as yet inexplicable to us, sportiness in a motorcycle has been stylistically idealized by abbreviated and mostly useless nonstructual parts. Given this theme, the Guzzi is certainly not sporty. As mentioned before, the tank is large and rotund. But its volume gives the bike a 200-plus mile cruising range. The fenders are very wide and deep and cover a lot of the tires' circumference. We rode the 850 about 60 miles on wet roads, during stop-and-go showers: those non-sporty fenders kept nearly all the wheel

The Ambassador is completely prepared as it comes in its crate. Wheels, for example, are so well balanced that they are dead smooth at 100 mph.

Instrument layout is easy to read, but indicator light patterns must be memorized. Key operates ignition, starter, and fork lock.

Dell'Orto carbs with 29mm venturis breathe through dry treated paper filter element. At rear is huge 30 amp/hour battery for starter.

Big .5 bhp starter motor complains but does job faultlessly. Rocker-type shift lever catches on boot heel when up-shifting.

New rear drive casting has a sediment sump to allow metallic particles a place to escape churning gear-oil.

Detailed finish of stodgily-styled components is perfect. Guard bar protects cylinders and graceful exhaust pipes.

spray off us. And that big, long, funny-looking, lumpily-contoured seat luxuriously supported one of our staff members and his wife on a day-long tour of Long Island that was the first long passenger ride that the lady had ever enjoyed. The tool boxes are just that: boxes. They are about three-inches deep and have lazily rounded corners. The shape is that of a canned-ham tin: kind of a 3-D triangle that was blown up with an air hose. Stamped steel side covers, just under the front of the seat, hide the huge battery (required by the electric starter), and the air cleaner. In shape, these covers are slight trapezoids, not far from being square. The only thing that keeps them from being nice-looking is the placement of a row of imitation breathing louvers over the portion which covers the air cleaners. Even the chrome plated pads on the sides of the fuel tank are entirely functional: They keep the rider's knees from rubbing the paint off the tank. Black rubber pads wouldn't look as good and would eventually deteriorate.

The headlight is half-egg-shaped and is supported by very plain stamped ears which attach to the top legs of very massive-looking forks. The tires are fat 4.00x18 Pirelli Supersport Universals and are mounted on Borrani Record aluminum rims. A golden eagle, like the one in the Sesame Street cartoon, flies above the logo on the tank. The net visual effect of the layout and styling is one of classical-funky, ageless, rich, utilitarian, graceful. In the way that an old Mercedes or Rolls sedan is grand, the Guzzi Eldorado is a Grand Motorcycle.

Sitting on the saddle again, we gaze at the stamped mini automotive instrument panel. There's an electric tachometer on the left, opposite a mechanical speedometer. Between these two instruments is the ignition switch and a quartet of indicator lights. A little black plastic hinged cover keeps water and grit out of the key hole when not in use. The key controls power to the ignition and starter only, but serves additionally to control a mechanical fork lock. With the key rotated to its most counter-clockwise position and/or removed, the forks are locked to the left. The first clockwise position unlocks the forks, but all power is off. Another twist of the key lights the little red indicator lamp to tell you the ignition is on and the battery is discharging. If the gearbox is in the neutral position, an orange lamp also lights. If the engine is cold, a lever up near the throttle grip is pulled clockwise to actuate the carburetion enriching system. This system is new to the Guzzi, and is a vast improvement over the pseudo-accelerator pump on last year's 750. With the throttle off, a touch on the dime-sized button down under the enricher lever actuates the electric starter. The starter groans its complaint at being summoned, but twists the engine over, *cha-glunk, cha-glunk*. It always started on the second *cha-glunk*, hot or cold. If we tried to start off before allowing the engine to warm for three minutes, the bike would hiss and pop its rebellion. Upon warming, the lumpy exhaust note tells you it's time to shut off the richener. A smooth but irregular idle develops that sounds exactly like that of a 74 H-D. It's a very deep, hollowish sound which has a soul stirring cadence, much like the drum lead to *Song Of India*. Without exception, people we encountered during the test liked the sound and said so without being asked or prompted. The rhythm of the sound is common to all V-twins and is caused by the irregular placement of the cylinders, and consequent irregular firing. In the case of the Guzzi, the left cylinder fires 270 degrees after the right, then there is a long pause while the crankshaft rotates 450 degrees and both cylinders breathe for another power cycle.

Sitting there idling, there is a noticeable sideways reaction of the bike to the crankshaft firing impulses. Since the crankshaft runs front-to-back (transverse), reaction to engine torque tries to rotate the whole bike like a goat on a barbeque spit. Just like a car. (On conventional bikes, where the crankshaft runs across the frame, torque reaction makes the rear end of the bike squat.)

An easy six pound pull engages the clutch and a push back with your boot on the rear pedal of the rocker-type gear shift lever engages low gear. On last year's test bike low gear engagement was accompanied with a resounding clunk. The clunk is gone, but at times the engaging dogs inside the gearbox butt instead of meshing. The result is that you push down on the pedal, and you feel something move not quite far enough. Sure enough, you let the clutch out and nothing happens. The answer is to roll the bike backward or forward a few inches to let things line up. Or just let the clutch out gently while maintaining pressure on the pedal.

Clutch action when moving off was very abrupt and jerky at first. We would let the lever out slowly, ever so slowly, and then get to a point where the bike would give a sudden lurch forward that would be very unnerving to our passenger. The clutch action improved somewhat after a couple hundred miles, but it never got anywhere near being smooth.

Once moving, even very slowly, all feeling of the torque reaction vanishes, and the bike becomes 550 pounds of rolling grace. It feels like a much lighter machine and responds quickly to light efforts applied against the handlebars. You can idle along at an easy walking pace and there is absolutely no jerking or roughness. It feels like there are fluid couplings and shock absorbers all through the drive train.

Slowly opening the throttle from the snail's pace causes the Eldorado to accelerate very briskly and with steam-engine smoothness. The bike loves to rev and is amazingly smooth at its 6500 rpm red line. Pushing down on the toe end of the shift rocker catches second gear with only a hint of the resounding clunk in last year's gearbox. Because of the huge amount of mass in the Guzzi's engine shafts and clutch, it takes a moment for the revs to fall to the appropriate level when you shift gears. If you do wait just that split second, the shifts are all velvet-smooth. Speed shifting produces a torque-reaction lurch that is more a novelty than an annoyance.

Moving out onto the interstate lets you know what the Eldorado is really all about: solid, high-speed, vibrationless comfort. That machine eats miles like no other mo-

Double dual-leading-shoe brake is operated through this adjustable cable anchoring device. Rubber boots keep out water and grit.

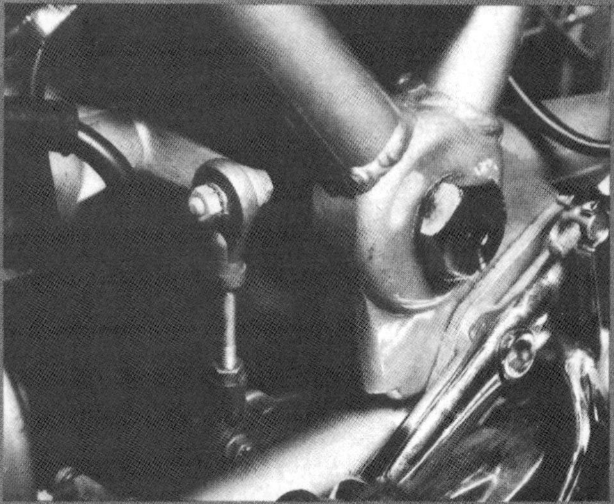

Frame tubes tie into outboard of swingarm pivot to prevent deflection under cornering loads. Heim joint is shift linkage link.

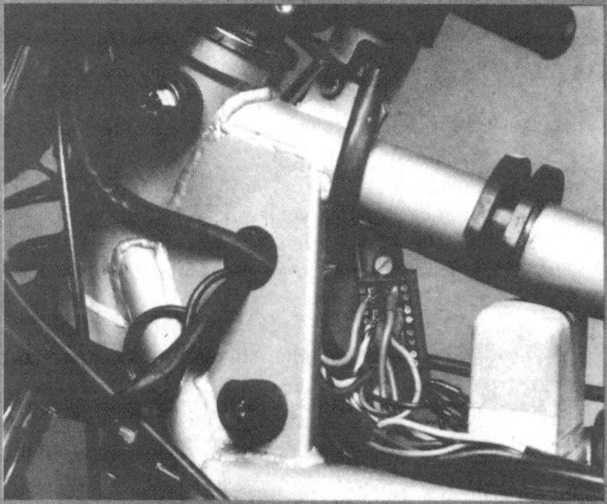

Heavily gusseted, hand welded frame is incredibly rigid. Modularized electrics are tucked away into protected areas.

Top frame tube trio tucks down between cylinders to keep overall height of Sport as low as possible.

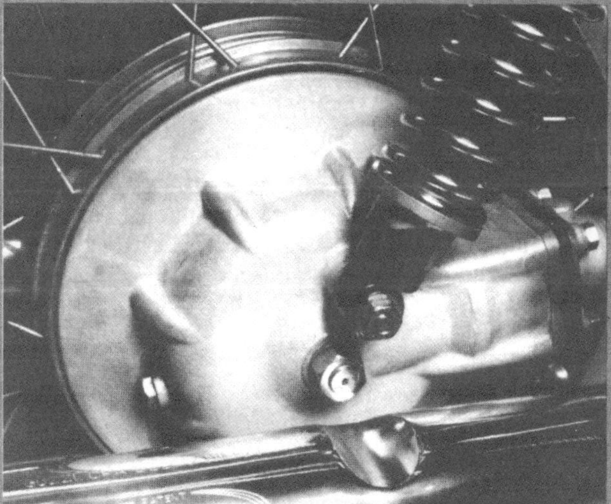

Notched mufflers allow maximum travel of Koni controlled rear suspension, while keeping mufflers high for cornering clearance.

Dual front brake should have more stopping power for experienced riders. Racing-type front forks work perfectly.

torcycle. Side winds and pressure front blasts from passing trucks don't faze it in the least. Ripply pavement and chuck holes go almost unnoticed by your backbone. And there's incredible cruising power. All last year's 750 needed was an extra transmission speed to provide an overdrive situation for cruising. What Guzzi gave its customers was the extra ratio and a lot more torque to boot. The extra 100cc obtained from lengthening the stroke 8mm (.314-inch) gives the bike a lot more urge. The machine cruises absolutely effortlessly at 80 mph and gets there in a hurry, if you want to. At the drag strip, the 850 ran 94.50 mph in 14.04 seconds: that's as quick as any Triumph Bonneville we've ever tested.

In more congested traffic, especially on winding, curvy roads, a few things gave us a little grief. The turn signal switch is located near the throttle grip, but just far enough away so that you sometimes have to take your left hand off the grip and reach over to turn the switch on or off. It's especially annoying when a hole opens in the traffic, and you can't signal quickly enough to take advantage of it. And, again this year, cornering clearance is rather limited. It is easy to scrape the center stand when leaning into a turn in either direction. And getting stopped is still a hassle: the brakes are marginal at best.

In normal turnpike cruising, our 850 averaged 42.6 mpg, and gave 36.2 mpg around town.

Throughout our test, the big V850 suffered absolutely no mechanical failures and did not leak a drop of oil. We had to replace one of the steel-shrouded spark plug caps that we accidently crushed against the side of our van when we picked the bikes up at the U.S. distributor, Premier Motor Corp., in Hasbrouck Heights, New Jersey. We did remove the tools from one of the locked tool boxes a couple of times to check the oil levels in the engine, gearbox, and rear drive during trips, but that was the extent of trip maintenance. No oil replenishment was required during any of our trips.

The only thing we did to the bike in our shop was to check the tappet clearance. It was spot on. We rode the bike and enjoyed it immensely, while taking in a lot of scenery. Our Eldorado was reassuring and inobtrusive at the same time. We simply felt that we could depend on it, no matter what.

In all, the Moto Guzzi V850 Eldorado is the most comfortable and least tiring tourer on the market. With better brakes and more cornering clearance, it could be perfect.

We had seen pictures of the V750 Sport in the European papers for over a year. The prototypes for the Sport were largely exercises of resentment on the part of Guzzi's design and development engineers. They

were tired of the new image, comprised of staid conservatism, that they projected with producing only the Ambassador.

There are a lot of the same people in Guzzi's offices and shops that were there when World Championship road racing bikes were being built. After all, they were still Italian, and the emotional reward derived from building a superlative, if stodgy-looking, touring bike could only sustain them for so long.

The Guzzi engineers placed the engine-transmission outline on the drafting paper and drew in the lowest, strongest frame they could within the bounds of lightness and human anatomy. Eighteen-inch wheels were placed at each end, in the right relationship to the frame, to give racing ground clearance. The Italian technicians then went to work on the engine to remove as much weight as possible. First to go was the huge automotive-type 300-watt generator from the crotch between the cylinders, and it was replaced with a small, compact unit that fits right on the front of the crankshaft. A much smaller Spanish-Bosch starter motor replaces the Ambassador's .7 bhp model, but only spots it .3 bhp. The touring model's massive flywheel/ring gear/dual-plate clutch assembly was lightened drastically.

The engine still looks practically the same as it did: like someone had taken an air-cooled V-eight automobile engine and band-sawed off the front six cylinders. The crankshaft, camshaft, transmission, and clutch layout is the same as such a car would have. The frame tucks so closely to the crankcase and between the cylinders that the bottom frame tubes had to be made removable, so the engine could be installed and replaced.

To create a more sporting power curve, the breathing components were replaced with ones which are more efficient at slightly higher revs. The same type square-slide Dell'Orto carbs are a millimeter larger than those on the tourers. Three-ring pistons with 9.8:1 compression ratio replace the stock 9.2:1 c.r. pistons which have an additional oil control ring at the bottom of the skirt. Of course, the camshaft pushes the valves open farther and keeps them open longer. The Sport's mufflers are larger in volume, though not noisier than the tourer's, and there are dual cross-over tubes so that both cylinders share both mufflers during scavenging. Closer-spaced ratios in the gearbox with a higher rear-end ratio allow useful advantage of the peakier engine. A complete dual point, condensor, and coil ignition system replace the single component distributor setup.

The suspension components have a racing look that gives warning of the Sport's capabilities. Up front is a very Ceriani-appearing set of forks that bear the Moto Guzzi logo. The bare, chrome plated stanchion tubes and aluminum sliders and triple clamps hold the same geometry as the touring bike. Koni multi-adjustable shocks control the action of the rear end.

A completely new set of dual double-leading shoe brakes are up front, while the standard rear unit has been converted to double-leading-shoe.

The long, low, sleek looks of the Sport are very sporty, in the very most Italian sense of the word. The frame is painted silver and an angular, 4-gallon gas tank nestles right down between the cylinders. A quartz-iodide Marelli headlight in an abbreviated shell fairly burns a hole in the road at night. Though less posh than the tourer's seat, the one on the Sport is beautifully soft and comfortable. The polished alloy fenders are short and narrow: good for keeping stones from hitting you in the eye, and that's about it.

Very special Michelin high-speed tires and Borrani Record rims complete the visual visceral effect of the Sport.

As a well developed and limited production personal superbike, the Sport is incredibly fine. We took it to Bridgehampton Raceway and thrashed it for a full afternoon of racing-speed exhilaration. The machine feels and handles like it weighed 200 pounds less than its 500-pound curb weight. It will corner at road-narrowing (in the mind's eye of the rider) racing speeds without dragging anything other than the tip of the footrest rubber. You can't exactly flick it around like a lightweight, but it responds a lot more quickly, with less effort from the rider, than any other machine of its type that we have ridden. The red line for the Sport is 7500 rpm, and we had it at 8500 several times. Even at those revs, the engine, indeed the entire motorcycle, is absolutely smooth. Uncannily so. We're accustomed to bikes that are fast being accompanied by a lot of noise and vibration.

The clutch is soft and positive in operation and the right-mounted shift lever has a short, crisp throw. We never missed a shift on the Sport.

The front wheel brake is built so that you can pull on the lever very hard and get strong, but not tire-howling, force. The rear unit is quite the opposite. You have to be careful not to lock the wheel during braking. We had to practice quite a bit to get the right combination of front and rear brake lever pressures, coupled with well-synchronized down shifts, to get the most effective slowing at the corners.

Riding the Sport on the street was no less satisfying than on the track, but only on open, traffic-free roads. The engine pulls smoothly up to 4000 rpm, then the cam timing really gets efficient. The power gets much stronger and from there to 7500 rpm in any gear, the bike accelerates voraciously. In the same way it is frustrating to drive a big-engined Corvette on the street, it is difficult to make yourself hold the Sport within legal bounds. It feels like it wants to charge.

The only thing that spoils the street manners of the Sport is the absence of an air filter of any sort. The carburetor bells open into a rubber boot that points down to, and almost touches, the crankcase. Not only is the engine susceptible to swallowing grit that might scar its chrome cylinder bores, but a lot of noise also gets out to bother the rider at cruising speeds. We'd guess that the production batch of Sports will have filters.

For riding around on the street, the short clip-on bars can be raised and pivoted.

Perhaps the Sport is an introduction to a new kind of ultra-performance motorcycling: one without the aggravation of a greasy, clanking engine. For the 300 or so lucky souls who happen to have an extra $2500 in their jeans this year, we do not know of a more potentially satisfying special edition to spend it on. ◉

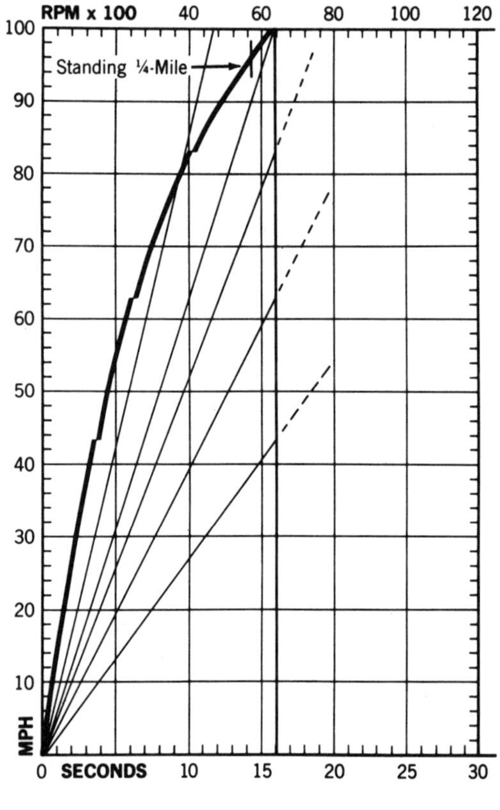

MOTO-GUZZI ELDORADO 850

Price, suggested retail	East Coast, POE $1985
Tire, front	4.00 in. x 18 in. Pirelli: Universal
rear	4.00 in. x 18 in. Supersport
Brake, front	8.7 in. x 1.7 in.
rear	8.7 in. x 1.7 in.
Brake swept area	93 sq. in.
Engine type	OHV V-Twin
Bore and strike	3.26 in. x 3.07 in., 83mm x 78mm
Piston displacement	51.5 cu. in., 844.05cc
Compression ratio	9.2:1
Carburetion	2: 29mm; VHB 29 Dell'Orto
Air filtration	Micronic treated paper
Ignition	Battery, single coil, distributor
Bhp @ rpm	64 @ 6500 rpm
Mph/1000 rpm, top gear	17.8
Fuel capacity	5.8 gal.
Oil capacity	6.5 pints
Lighting	12v, 300 watts
Battery	12v, 32ah
Gear ratios, overall	(1) 11.42 (2) 7.92 (3) 5.98 (4) 4.96 (5) 4.28
Wheelbase	58 in.
Seat height	31 in., with rider, shocks soft position
Ground clearance	6 in., with rider, shocks soft position
Curb weight	560 lbs., with ½-tank of gas
Test weight	720 lbs., with rider
Instruments	Speedometer-odometer, tachometer, warning lights
Standing start ¼ mile	14.04 seconds 94.65 mph
Top speed	116 mph

GENTLE GIANT

Moto-Guzzi V7 Sports . . . the most expensive bike on the British Market

WHAT is a "Superbike"? Is it to be judged by price? Engine capacity? Glamour? Performance? Whichever applies, the Moto-Guzzi V7 Sports qualifies as a Superbike with no reservations at all. It is just about the most expensive readily available motorcycle on the home market, at 750 c.c. it is in the same league as most of its rivals, it comes from a factory with a reputation second to none for quality and character and, with a claimed top speed of 128 m.p.h., it numbers among the quickest.

Our test machine, on loan from Rivetts of Leytonstone, is one of the few, is perhaps the only one, of its type in the country. It is a crowd puller. Every time we left the machine we came back to find people looking at it. We could easily distinguish the *active* motorcyclists for they would tell us that this was the one from Rivetts' window! The majority, of course, had no such inside knowledge. Sometime, maybe a long time ago, they had been motorcyclists. The Moto-Guzzi brought to the surface all the dormant enthusiasm that never really leaves an ex-motorcyclist. Questions were asked, eyes would light up: who knows, perhaps one or two of them rushed away to place an order at Rivetts! The Moto-Guzzi was like that. It looked interesting. It was not a machine that would be difficult to find in a crowded bike park. Lime green tank, red frame and stainless steel mudguards are not the combination that I would have chosen for my Moto-Guzzi —but, by gosh, you could certainly see it coming!

Some months ago we wrote about the V7 Special from the same factory. The same basic layout applies to the V7 Sports: transverse vee-twin engine, engine-speed clutch, shaft drive. There the similarity ends. The Special, a charming machine, was also rather portly and its performance, although reasonable, was a little mild. The sports version is considerably changed. The engine has had a power boost from 60 to 70 claimed b.h.p. and has undergone a drastic slimming course, with the weight reduced to a much more manageable 454 lb (even Moto-Guzzi must be embarrassed by the weight of the V7 Special for it is the only machine in their sales leaflet for which a weight is not mentioned!)

How has all this weight (something like 75 lb) been saved? Now that is a good question. The engine unit has much the same bits and pieces, for both machines: electric starter, generator, alloy heads and barrels, Del 'Orto carburettors. The gearbox on our test machine had five speeds as opposed to four on the "cooking" version so I doubt there was much weight saved there. The frame? The frame of the V7 Sports contributes a lot to the excellent handling. It is a full cradle, holding the engine snugly in its grip but willing to release easily, for the right-hand bottom tube can be easily unbolted to facilitate engine removal. A fine frame, but is any weight saved? Not much, we would have thought, for the Special has much less of a frame, the engine being part of the structure. Front forks? Yes, a little there, for the Sports has Ceriani type as opposed to old-style on the Special. The battery? Yes; the massive car-type battery on the Special has given way to a smaller 12 volt one. It still turns the engine over with ease. Wheels? Tyres of 4.00x18in section, front and rear, have given way to 3.25 x 18in front and 3.50 x 18in rear, giving the machine a less bulky look and, of course, requiring less effort when it comes to manoeuvring at low speed. A four-shoe front brake, two twin-leading-shoe brakes back to back, requiring hefty cables and controls, and the twin-leading-shoe rear certainly do not combine to save weight. We finish with the problem solved, not by any drastic and obvious attack on any one department, but by small and careful savings wherever possible . . . things like lighter foot controls, front mudguard, headlamp shell, handlebars. Does such a weight saving and performance boost make up for the cost of the Moto-Guzzi V7 Sports? The V7 GT 850, which is now on sale as the "touring" model, sells at more than £1,100 over here; the V7 Sports has a whopping £1,350 price tag on it, enough to frighten off all but the most dedicated connoisseur. A fairer comparison would be to consider the prices in Italy, for by the time the Guzzis reach the British market they have a price inflated by import duty, purchase tax and handling charges, so that £350 could be halved at the outset.

From the moment one swings a leg over this dazzling green and red monster it's plain that this is a *man's* machine. One is immediately aware of the massive engine, the rocker covers so near to one's legs that moving forward a few inches brings the knees into contact with the heads. The choice of handlebar positions is considerable, with the clip on handlebars each held to the fork leg by clamps retained by socket bolts. The rider has a choice of 4in up and down movement with considerable backward and forward adjustment as well. These bars *must* fit. The dualseat is long and firm but very comfortable, the kind of seat that is not going to feel like a plank after a day sitting on it. The choice of control extends to the gear-change and brake for it is a simple matter to move to the gear-change from the standard righthand side to the left. Many prospective buyers, I suppose, are put off a machine that otherwise appeals by the gear-change being on the "wrong" side.

There are two choices when starting the machine; either the key is turned, car style, or, once the ignition has been activated, the button by the right thumb will perform the same task. A vee-twin definitely has a note and character all of its own, of course, a lop-sided sort of tick-over, smooth to feel, lumpy to hear. The Guzzi was happy to plonk away with its slightly uneven beat at 1,000 r.p.m. for ever, with mechanical noise at this, or any other speed, being almost nil. The exhaust was very subdued too. A quick flick of the throttle brings a lurch from the motor as the torque reaction rears its head. It is no more of a problem on this machine than on any other with flat- or vee-twin layout.

The engine-speed clutch, with twin disc clutch plates, was reasonably light and very positive. One knew exactly when the power was going to be taken up. The five-speed gearbox was one of the slickest we have tried on a machine with an engine-speed clutch. Provided that one followed standard practice for such a system, allowing a moment's dwell after closing the throttle for upward changes and giving a roll of the throttle as the clutch was disengaged on downward changes, the change was light and completely inaudible. Rush it and the usual *clunk* would occur. Choice of gears is "touring" with a fair jump from ratio to ratio. Top gear is very high (4.09 to 1). Bottom allowed the machine to be trickled along with the clutch home and the speedometer needle barely moving on the dial with a touch of throttle bringing the machine leaping forward, the chainless drive transmitting the power to the rear wheel without trace of a jerk. So smooth was the power delivery that the V7 would accelerate smoothly from 10 m.p.h. in top gear. Engagement of bottom gear was always silent, though selection of neutral was a slightly delicate matter, with little movement of the pedal at this point. A yellow light indicated neutral and was right 99 times out of 100. The 100th time was the one that worried us and we soon learned to check first. It had caught a previous tester out. He had pressed the button with the bike in gear, landed in the road and removed the centre-stand lug. We might have suffered the same fate had he not warned us!

There can be few more comfortable machines on the road today: comfort combined with solidarity. With the massive engine unit right there in front one has a sense of well being, as though bulk alone might discourage impatient motorists from taking liberties. We think it *did* too. The 4½-gallon fuel tank, finished in a most attractive lime green (the red frame disagreed with it), is sculptured for the knees. An excellent shape but once one introduces knee shapes into tanks the onus is on the rider's knees to fit. Mine did, most of the time, but occasionally I slid forward on the seat and found the edge beginning to intrude on my comfort. One is not particularly conscious of the Guzzi's 450 lb, even when trickling through heavy traffic; in these conditions the only thing to be careful of is that one doesn't trap the left thumb between the handlebar and tank. It is a surprisingly good town bike (in our book

A motorcycle of rare character and performance

any machine that is not a good town bike loses a good deal of its appeal for, like it or not, we still have to ride motorcycles in towns). So smooth was power take up that with the tachometer registering a 1,000 r.p.m. tickover one had only to release the clutch to glide forward with utmost ease.

Perhaps the brakes are better suited for the open road, for the double 2ls front, operated via twin cables, required considerable hand pressure to give of its best and that was tiring after a while. The rear brake—2ls again—was so good that one was constantly fighting a tendency to let it do all the work. Tyres are Michelin. When I had my first motorcycle I fitted a Michelin and it endeared itself to me as a most supple tyre. Happily we never needed to take a tyre off the Guzzi to investigate whether Michelins still have this quality. No complaints from me about its other properties. It had a deep tread, very deep for a front tyre, gave good handling, wet or dry, and generally acquitted itself well. Michelin, I learn, have been collaborating with Moto-Guzzi in the production of high-speed tyres.

Performance? A claimed 128 m.p.h. Yes, well, you see while we are prepared to go to great lengths to satisfy our readers' thirst for facts and figures, finding a road long enough and safe enough in this country to attempt that sort of speed is not easy. What we are trying to say is that we can't be *sure* the V7 Sports will reach 128 m.p.h.—but we can be sure that it will handsomely exceed 110 m.p.h. Will that do? Top gear is so high that it would take a very long road to find the Guzzi's absolute limit. To get best performance one needed to hold on to fourth gear until the k.p.h. speedometer indicated just over 170 k.p.h. (about 106 m.p.h.). Engaging top gear at this speed produced a continuous flow of performance. Changing up at anything much less produced a decided lull in the proceedings. The legal speed limit could be handsomely exceeded in any gear after bottom, and for real performance the first three gears would amply suffice in this country. It is too much like hard work for my taste, though, and I found there was a great deal more pleasure to be had in whoofling along at 70 m.p.h. in top with the tachometer showing 4,000 r.p.m. At 50 in top one really did feel that the motor was firing at the proverbial every other lamp-post! The handling of the Guzzi was impeccable. It would not, to be sure, bend swing like some of today's outstanding middleweights, nor should one expect it to, but it pounded through almost any going that one could find with complete confidence. With a 4-cwt bike I tend to ask, rather than order, it to go where I want! But this should not be taken as evidence of a wrestling match, far from it. A hydraulic steering damper is fitted that can be brought into action by a knob in the traditional steering damper position. We were unable to tell whether it was connected most of the time and the only occasion when we would think it useful is at very high speed over an extremely bumpy road. The Moto-Guzzi's halo slips a little at night. The 45/45w headlight, even with a 12 volt generator behind it, was not really as good as we expected. Far too much light was lost outside the essential area. Twin horns are fitted; they were reasonable but one felt that they ought to have had a really *penetrating* note. And the arrangement of the switches is not really up to the style of the rest of the machine. All the switches are in one unit near, but not near enough, to the left thumb. On top is the horn and dipswitch with the on/off light switch at the side. Outstanding was the ignition key, which doubled as a steering lock. Pressure as one removed the key clicked

Continued on page 172

Dualseat up reveals fuses, expanse of tyre

Clip-ons with wide range of adjustment

Cross pipes for exhaust under crankcase

Ignition switch (starter) and instruments

Twin front brake: four leading shoes

Rear drive; multi-setting shockers

MOTORCYCLIST TEST:

MOTO GUZZI ELDORADO 850

AN IMPROVED 750 AMBASSADOR. WHAT MORE COULD YOU WANT?

BY DALE BOLLER

Another measly 100cc really helps. You can feel the 850 belt you in the tail end with a handful of throttle whereas the 750 eases up and gives you a little pat, responding slow and sure as a touring bike should. The 850 snaps. Passing is safer, you can surprise a ration of "superbikes" with near equal performance and none of the renowned smoothness is lost. The whole 850 is a pleasant improvement over the already-superb Ambassador, making it a stronger-than-ever consideration for touring enthusiasts.

Those riders interested in the Guzzi are a unique segment of the touring market. They can't relate to the Harley-Davidson Electra Glide or its image and they won't push the big sport bikes (Honda Four, Triumph Trident, etc.) into touring service. These alternatives aren't in the tradition of ultimate highway cruising — you prefer to be astride a prestigious hand-crafted example of European precision, exuding steadfast pride in ownership all the way down the road. You have great respect for your motorcycle's technical prowess and reliability. You're riding a machine carefully thought out for you and you know who you are. You are not about to accept substitutes.

This proud rider, and there are many of them, has been a BMW owner for 40 years, until 1967 when Guzzi's 700cc V7 first came on the scene and gave him a choice. For this breed the Moto Guzzi is built and its purpose is clear — to travel long distances without protest even if laden with camping gear and passenger. There is no compromise in cost or weight. Yet this mile-eater is versatile enough to serve dozens of police departments in tortuous traffic within confining city limits.

There is no doubt the extra 100cc was added to make the Guzzi better compete with the giant 1200cc Harley in police use. Any police machine must carry about 70 pounds of added equipment, which sops up power and performance at an alarming rate. With the 850's 4½ more hp over the 750 and its considerably greater torque (torque, not horsepower is the prime mover of weight) the 850 is compensating for a great percentage of the police load and consequently performs right in line with a Harley, if not better. About 100cc were needed to achieve this par in performance, so that's why we have an 850 Guzzi. Mechanical considerations did not limit size because Guzzi

About the only time you'll see this motor apart is in a drawing. The real thing rarely needs to be pulled down.

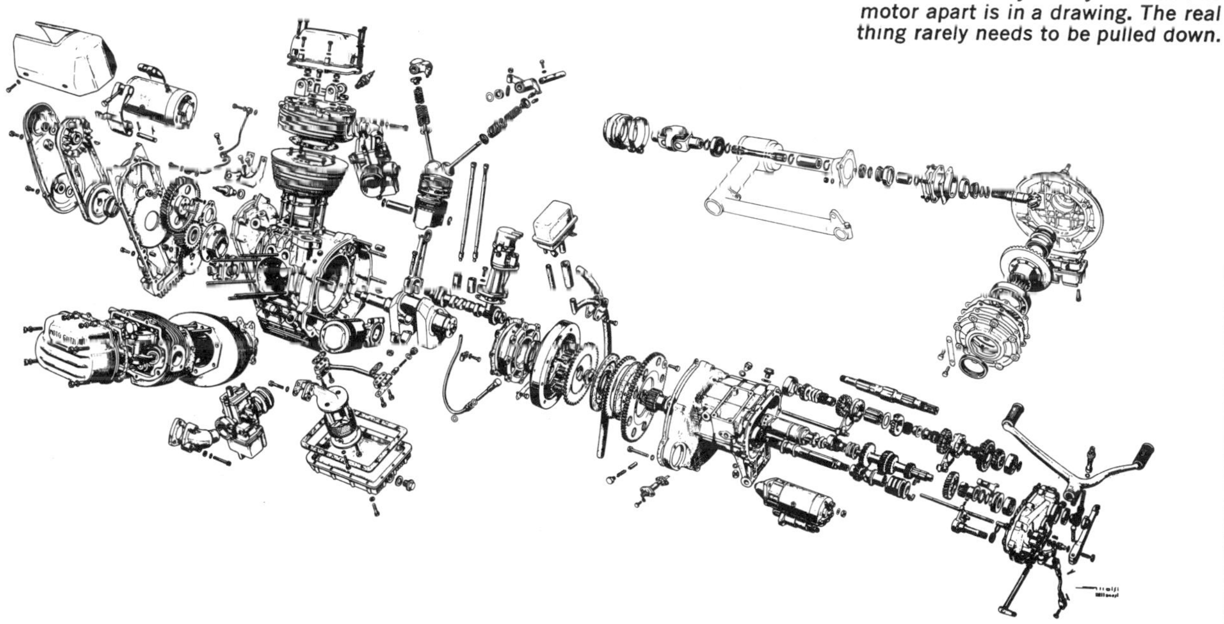

• Overall styling succeeds admirably considering huge bulk of components, though more compact headlight and rounded sidecovers would improve flow of lines. Big valance on fenders halts rain-fling from tires and is appreciated only when experienced. The touring crowd knows. Centerstand leverage enables anyone to hoist $\frac{1}{4}$-ton bike easily. • Driveshaft passes right through swing arm to rear hub which has a new finned sludge trap visible below the muffler. Locking side container marked "850" is an empty storage box; its twin on left side contains the tool kit. Note large filler cap nuts on both the trans and rear hub. Similar drain and level plugs aid and hasten oil changing.

• Since our test bike was one of only two 850's in America and West Coast Distributor Bob Blair was curious about the new five-speed gearbox, he decided to tear it down while we watched. His competent mechanics had the entire transmission on the bench in just 27 minutes! Blair says the whole machine is simple and fast to service, and after seeing a complete tune-up and oil change in less than an hour, plus the trans removal, we're sure he's absolutely right. A two-plate dry clutch is enclosed within flywheel and external gear teeth are for starter engagement. • Centralized wiring and fuse block inside the headlight are equipped to take all police accessories so civilian touring rider gets a free set of spare fuses! • At right Art Director Paul Halesworth shows you can lean far enough for touring purposes without scraping the centerstand.

took the hard way out and stroked their motor for the increase anyway, and if they needed even more displacement, they could have punched the bore in addition. Moto Guzzi wants police business and they're doing everything necessary to get it. Let's hope they never stop trying, because we end up with a better bike for our more mundane civilian pleasures.

Moto Guzzis are built on the shores of Lake Como in Italy where the factory began building motorcycles in 1921. About 1200 workers have a hand in the operation. They build it big and strong and their quality control is most impressive. We examine every nut, bolt and screw on a test bike before entrusting our life to its care, and invariably, regardless of brand, three or four fasteners aren't snug; or the headlight isn't aimed properly or the brakes drag or a cable is binding. Dealer setup you say. On some brands maybe so. But the Guzzi comes set up in the crate. All the dealer does is tighten down the bars and charge the battery. On our test machine *everything was perfect,* and this rarely ever happens. Also we weren't using Phillips head screwdrivers to check the fasteners. We used hex keys, because the best socket head bolts are employed throughout. Take a close look sometime; the bike is built like a battleship, but Guzzi owners wouldn't have it any other way.

To test a Guzzi you must go on a trip. We took off on a pleasant springtime Saturday and rolled through 710 miles of varying terrain before Sunday's sun had set, or about 350 miles a day which is just about the maximum if you expect to enjoy the scenery or have time for possible roadside attractions. Later picture taking expeditions and two weeks of riding in the city reinforced our findings on the trip and confirmed all aspects of the Guzzi's purpose and demeanor. The machine is brilliantly tailored for its job, and performs it with imperturbable dedication and ease. Within the confines of its staid touring purpose, absolutely nothing is faultable — except tricky clutch action which requires extremely deft release of the handlebar lever to pull away from a standing stop without spilling the soup. We have a few other little picky criticisms, but the clutch is our only major gripe.

Climbing aboard and settling deep into the saddle of all saddles, one suddenly realizes it's easy to touch the ground, an unexpected relief for many who may have viewed the ominous bulk with apprehension. Controls are well placed and sufficient in number to keep tabs on vital functions — oil pressure, generator, high beam and neutral lights flanked by a tach and speedo unfortunately unequipped with a trip mileage indicator. This is a must for a touring bike. Also the neutral light frequently registered when we were still slightly in gear, but since feeling from the gearbox through the foot was so positive, we usually knew when the light was lying. Any dealer can correct the whole problem by careful realignment of the neutral light pick-up in the gearbox, a no-sweat job they shouldn't charge you for. We'd also like to see a new turn signal switch since the present one requires removal of most of your hand from the throttle to activate and can easily turn on one signal while turning off the other by passing through the center off position.

Foot controls are not limited in access due to protruding cylinders or engine cases as on some machines, and are interchangeable to switch the brake to the left and shift to the right with an optional factory kit. A stock 850 has left-side shifting with an up-for-low pattern controlled by a heel-and-toe shifter. As with any drivetrain made up of splined shafts, there is a noticeable clunk when engaging gears with a heavy foot, due to every spline in the system bumping against its opposite bearing surface as a sudden load is applied or removed. The clunk is more irritating to the rider than harmful to the driveline and may be almost eliminated by slow, deliberate shifting, a technique any touring rider uses anyway to keep from jerking his passenger. A Guzzi owner must develop several delicate techniques to eliminate clunk and to smooth out the giant flywheel and touchy clutch. Anyone can ride a Guzzi, but those who ride one well are superior motorcyclists.

Heading through the city toward the open road, we see why the police feel no pain after 8 hours of stop and go. The bike barely reveals its weight in slow going (though there are more nimble machines) and the motor slugs along in third at 20 mph. It settles into a perfect idle at every red light. The clutch is easy to pull and the trans responds nicely to careful shifting. Neutral almost appears by itself at the next light. Don't wing the throttle while waiting — the carbs have accelerator pumps and may flood, and also you'll get quite a roll to the left as the whole bike leans from the engine's considerable torque. This torque, and a feeling of uneven motor pulses arising from 90° cylinder configuration, disappear completely once underway. Above 2500 rpm it's all glass, right up to our top speed indication of 108 mph in a semi-crouch, and especially in the 70-mph cruising range where the tach rests steadily at 4200 R's.

DIFFERENCES BETWEEN 750 & 850

	750	850
Bore & stroke	83 x 70mm	83 x 78mm
Compression	9:1	9.2:1
HP @ RPM	60 @ 6500	64.5 @ 6500
Gears	Four	Five
Weight, wet	539 lbs.	569 lbs.
Turn signals	No	Yes
Self-return choke	No	Yes
Price	$1753	$1989

- The 850 has new internal and external horizontal reinforcing webs on the crankcases.
- Clutch and gearshift linkage are rerouted.
- New five-speed gear cluster is housed in a redesigned transmission case with heavier bearings.
- The rear hub is a stronger casting with oil capacity increased from ½ pint to 1 pint and a new finned sludge trap on the bottom.

MOTO GUZZI 850 ELDORADO

Model • Eldorado 850
Suggested retail price • $1989
Engine type • 4/S OHV V-Twin
Displacement • 836cc
Bore and stroke • 83 x 78mm
Claimed HP @ RPM • 64.5 @ 6500
Compression • 9.2:1
Carburetion • Two 29mm Dell'Ortos
Ignition • Battery and coil, distributor timing
Electrics • Bosch generator, voltage regulator
Battery • 12V Varta
Starting system • Electric
Clutch • Twin dry plates
Transmission • Five speeds
Shift pattern • Left, up for low
Start in any gear • No
Final drive • Shaft
Frame • Double cradle
Shocks • 3-way adjustable
Brakes • F, twin leading shoe; R, single leading shoe
Tires • F, 4.00x18 Pirelli; R, 4.00x18 Pirelli
Rims • Borrani Alloy
Wheelbase • 59 inches
Length • 85 inches
Seat height • 32.5 inches
Peg height • 12.5 inches
Peg to seat • 20 inches
Handlebar width • 30 inches
Ground clearance • 6 inches
Weight, wet • 569 pounds
Tank capacity • 6.1 gallons
Gas mileage • 50.3 mpg
Indicated top speed • 108 mph

We're on the open road now, home free in the all-new fifth gear, and there's nothing else you could ask for. Roll it on and she moves right out without hesitation, without great decibels of exhaust noise and without a wisp of pollution. Clamp the binders and the big freeway freight train sinks down on its forks and halts straight and hard, but in this era of the disc, why not make adequate braking superb braking? Maintain a sane speed in the mountains and handling is confident and sure; play racer and the centerstand scrapes before you're over far enough to qualify for the Novice heats. Only the centerstand limits overall handling prowess since suspension and geometry are up to much higher cornering speeds. But for touring purposes there is plenty of clearance.

• Gigantic crank rotates in equally hefty plain bearings pictured below it, the larger one being at the gearbox end. Note missing flange section at 5 o'clock on this bearing so correct installation position is assured. Rod bearings are alloy inserts, easy to machine, cheaper than rollers and good for 100,000 miles. Smart design.
• Layout of four-speed drivetrain from 750 helps clarify similar setup on five-speed 850. Spring on lower left shaft absorbs most drivetrain snatch, a common annoyance on many other machines without such devices.
• To right of drivetrain is clutch which needs a wider friction point.
• Scene below is what the Guzzi's all about. Open highway, 75 mph, end of the day and a contented enthusiast.

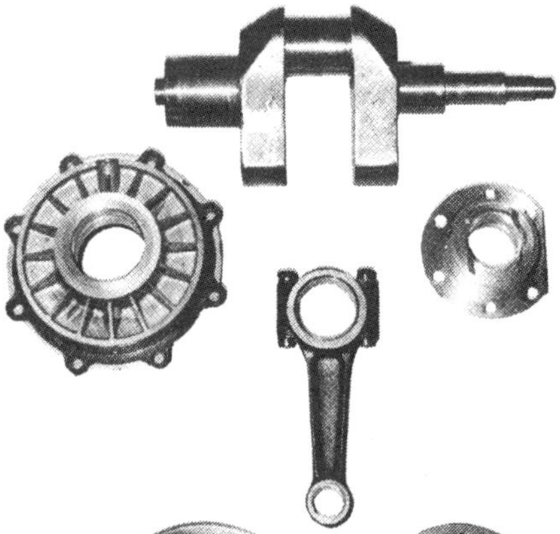

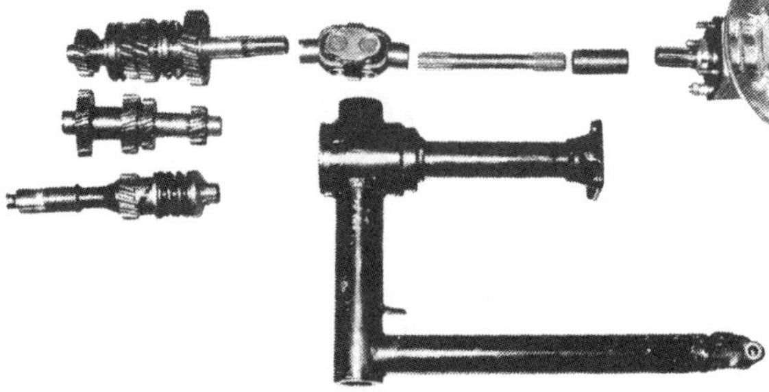

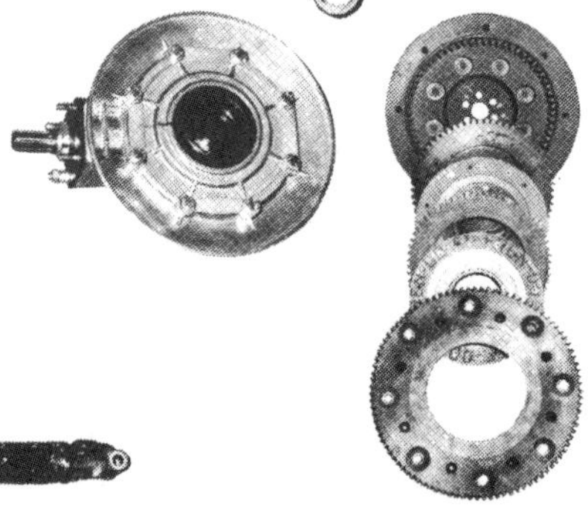

Forks and shocks provide a secure but stiff ride, insensitive to freeway seams. Suspension must be stiff to support the usual load of touring gear and the bike's own weight of 569 pounds with 6.1-gallon tank brimming. It's amazing that BMW builds a similar machine which scales just 458 pounds wet. The Guzzi's extra gallon-and-a-half tank capacity and its fifth gearbox speed account for 20 pounds, and its giant battery for an additional 15 stones over the German, but where is the other 76 pounds? Weight should not be a deterrent in your consideration of a Guzzi, but why lug extra around when it seemingly isn't necessary?

Weight may be an advantage as the day wears on for the Guzzi's solid and steady gait is easy on your body and the ride never gets too long. Remarkable mileage of 50.3 mpg combined with the big tank gives a safe cruising range of 300 miles — which you could cover in one sitting without great fatigue if your bladder capacity was more than mine.

What makes the 850 such a runner? It's the man in the saddle; he has enough confidence in the engine and its reliability to go anywhere in the world without so much as a spare set of plugs. Guzzi riders hover at 70-75 mph all day because they know their bike is built for endless highways and that it's well up to any touring task. Why? Conservative and proven engine design is the answer. Moto Guzzi chose the simple 90° transverse-mounted V-twin concept for three basic reasons: one, cooling is extremely effective; two, there is an inherent smoothness and only a minimal rocking couple from the offset nature of the rods; three, and most important, transverse mounting best facilitates shaft drive.

Components throughout the engine are overdesigned — sized or strengthened beyond their requirements — to achieve reliable service and long life. A one-piece forged crank rotates in huge automotive-style plain bearings good for 100,000 miles. More plain bearings, bathed in a constant flow of oil circulated by a tough gear-drive pump driven off the crankshaft, serve the rods. Alloy pistons have 9.2:1 compression while four rings prevent traces of oil from sneaking into the combustion chamber and escaping through the exhaust as smoke. A single gear-driven cam between the cylinders activates a standard valve train, pressure-lubricated under the rocker box covers by oil which drains down the pushrod tubes into the engine's copiously finned 3½-quart wet sump. An improved Bosch generator on both the 750 and 850 mounts on the top-front of the engine where a belt running in an adjustable pulley on the crankshaft provides constant spin. The automotive-type distributor with automatic advance and conservative 38° timing, angles to the surface between the generator and right cylinder from its drive point at the rear of the camshaft. A single coil tucked under the tank boosts battery voltage to spark the plugs. Right- and left-side 29mm Dell'Orto square-slide carburetors, with accelerator pumps which mechanically supply a burst of extra gas to the main jet with any full throttle twist, serve well in all ranges and are easily accessible from the saddle for idle or mixture changes.

Electric starting by a heavy-duty Bosch motor which gears up to the flywheel is aided by a handlebar choke control. No kickstarter is provided, and if that bothers you, you must also worry that there's no crank with your car. Starting is always a snap, but must be followed by a long warm-up (at least 3 minutes) when the engine is cold to prevent dying under load.

This new 850 motor, 8mm longer in stroke than the 750, produces noticeably more torque, stronger acceleration and almost 10 mph more top end from a horsepower rating of 64.5 at 6500 rpm. In a dramatic comparison with a BSA Rocket 3 concurrently under test at *Motorcyclist,* the Guzzi's lumbering torque enabled it to pull away from BSA's triple in 50-mph throttle roll-ons in fourth and fifth gear. Both were even in third gear combat and the BSA was able to jump the Guzzi about one bike length in second. A Moto Guzzi is definitely more potent than it appears.

How its considerable power is transmitted to the rear wheel is another interesting technical design. No less than four shafts and six gears are involved and the whole system is considered quieter, more reliable and more maintenance-free than chain drive. A giant flywheel enclosing the twin-plate dry clutch connects to a spring-loaded shock take-up shaft in the transmission which gears to the mainshaft. The shock shaft has a big job in taming the huge flywheel and does it well since lurch in shifting is considerably less than in the 750 BMW tested last month. From the mainshaft power is processed by one of five gears on the layshaft which aligns with the rear-wheel driveshaft and turns it through a universal joint mounted in heavy ball bearings. The driveshaft, located inside the swing-arm tube, is in two pieces joined by a sleeve so the front half can slide forward and back within the sleeve as the swing-arm moves. Final right-angle gearing is accomplished by a ring and pinion in the rear hub which is redesigned with stronger bearings and more oil capacity.

Combining all the inherent sophistication of the 750 with refinements designed into the 850 results in one fine motorcycle which is a pleasure to endorse as a touring machine. We would make these suggestions for improvement:

- Smooth out clutch action.
- Reduce overall weight.
- Fit more accessible turn signal switch.
- Consider a disc brake system.
- Eliminate scraping centerstand.
- Establish more dealers.
- Design the starter so it will operate in any gear.

The bike doesn't *need* these changes to be good. They'd just add a little whipped cream to the frosting. And to end on a even sweeter note, check the price — $1989. That's over $100 cheaper than a BMW and up to $500 less than a full-dress Harley. Kinda like getting your tax and license free if you buy a Moto Guzzi. •

SOMETHING TO DO WITH SANITY

REPORT AND PHOTOGRAPHY: DAVE MINTON

Y'KNOW, I sometimes think that journalists — the riding journalists that is, and there are a damn sight fewer of those than there should be — don't always understand exactly the national bike scene as they ought.

We ride around on all sorts of lovely machines that most fellows would hack off an arm for, and then criticise the thing because it twitches a bit over 100 mph after three hours down some sun-baked autostrada. It's all too easy to smugly re-adjust to the new situation, which, were it not for paid expenses, the borrowed and well-serviced machine would be entirely beyond our means. What I mean is that you take some keen biker — me for instance — and plant him in paradise and tell him to work. He (me) doesn't think. "Hey! This is great, and I'm a lucky guy but it's all a bit unreal so let's keep things in perspective." He (me) thinks, after a snap readjustment. "Hey, this 750 won't pull 110 for long without blueing its pipes and the front brake fades if you grab it more than three times in a row so it ain't all it's bulled up to be so I'm never going to buy one." All the while conveniently forgetting that the thing is monstrously unsuited to his real-life needs.

Life in Daytona was crazy. There was me, chunkin' around on a throbbing great Harley', borrowing MV fours and race

tuned 1,000 cc Honda Fours and Laverda threes driving 7-litre Pontiacs and mixing with people like Cal Rayborn and going for a swim before breakfast in the heated outdoor pool ten feet from my hotel room, yet at the same time supposed to keep a level head and work!

I was high, up there somewhere, on a different level to mere mortals, not out of my depth, but walking where others swam.

Then I met Tom Lester and he lent me, as well as his MV and Laverda, his Moto-Guzzi Falcon. It settled me down on my feet gently and firmly as only such a wholly practical bike can.

You don't know what a Falcon is? Like the Aermacchi (now Harley Davidson and available) and Benelli singles, 'Guzzi roadsters never caught on over here, although with all the caterwauling that went on as our singles' industry died I'm blessed if I know why. The Falcon is a tidied up version of the traditional old 'Guzzi mainstreamer now fast becoming a Continental classic among collectors.

It used to be red, a flat single with an outside flywheel, leading link front forks and damped at the back end with Hartford style friction pads. Lovely bike; ugly, oh so bull terrier ugly, but so proudly, satisfyingly so. The rest were wrong, it was right, and it knew it; you could see so, in every contrary line.

The Italian police used it, and the army, and farmers and small traders with their stalls towed behind them on the way to market used them. The big old single was the workhorse of Italy.

In time, though, even the army wanted something more conventionally appealing. 'Guzzi cleaned up the angular single, tucked the flywheel inside a casting, added teles and pivoted fork suspension and produced the Falcon.

Amidst the metalflake furore of Daytona, the Falcon came as a blessed symbol of reality and I appreciated its simple grace all the more. As I rode it, I thought, "Why all the razzmatazz about size and multis when something like this does the job a whole lot better for 90 per cent of the time?" And it does you know, at half the cost yet with far greater riding ease.

Like the 350 Harley' the Falcon has much more for it than just about any other machine type you care to consider. Low centre of gravity, light weight, simplicity of design and servicing requirements, and Italian roadholding and braking.

Only a few years ago, I would have rated our own machines as the finest handlers in the world, but now, with the Italian industry taking over as the leader of Europe, I would also put them as the makers of the machines most likely to offer a high-grade performance quality. Just think of them, Benelli, Ducati, Laverda, 'Guzzi, and all with superb handling characteristics.

I spent only a morning with the Falcon, but in that time realised that it was a machine that in all respects would remain beyond the ability of its rider to better. It felt so small after the ponderous Electra-Glide, the surging Laverda and the big Honda (I leave the MV out of it, but more of that later), that I required no breaking-in period. From the time I was riding, everything went well.

Even for a "mere" 500 single, it sits low on the ground — it can afford to with such an engine — so turning and bend swinging with fast direction changes following one on top of another could be carried out at speeds that are fast becoming unfamiliar to riders of high, wide and heavy modern machinery. But at the same time, because of the clean lines and available space, footrests sat placed higher, narrower and slightly more rearward than is usual on current machinery. Nothing racy, but right, very right.

A rev counter was fitted, although it proved entirely unnecessary due to the engine's relaxed power output, but it did make testing simpler. Whatever the makers might state about optimum power output, it's a wide spread below that, but because it was a single the engine was none too happy below 1,800 rpm in top gear. As this corresponded to only 32mph though, it could scarcely be grounds for complaint.

Through the gears, maximum speeds were obtained only at the expense of some vibration at high revs, which while not bad, contrasted acidly with the Falcon's otherwise smooth power flow. Nothing like a big Suzuki, or even a Honda Four — dammit, but those rustling monsters have got a lot to answer for when you think about it, eh? Forced us to completely revise our standards. Three years ago, the Falcon would have been worth raving over, but now it's simply a uniquely rational motorcycle. As with all big singles, what movement there was came less from resonance than the rhythm of big bits moving around. No, not vibration, merely a feeling of working metal.

Gear speeds resulted in maximums of: 1st (13.23:1 at 5,500 rpm) 31 mph; 2 (7.42 at 5,500) 51; 3, (5.51 at 5,500) 75; 4 (top, 4.12 at 5,000) 88 mph. The thing was choked, or at least it felt like it, although the carb was by no means undersize, at least by our standards, for the Italians do like big intakes. With all due respect to them, I am confident that our own singles of com-

OVER

SOMETHING TO DO WITH SANITY

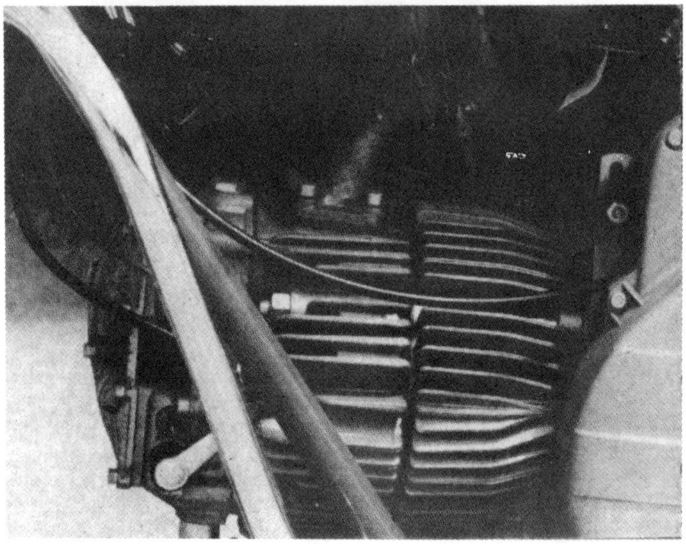

parable type were better equipped for the job of slogging, yet still offering a fair top end than theirs.

Like Ducati Sebring, another delightful single, the Falcon's performance feels definitely "withheld" somehow. It just has not the elastic guts of say an old Matchbox or Red Hunter. Same top end mind, nothing in it, but without the sleepy strength at rock bottom that was one of the most lovable traits of our singles.

It was too cammy, too compressiony and not flywheely enough. I stalled it treating it as I thought it should be. The makers have obviously gone to a lot of trouble to ensure efficient gas emission though, at least at top end. Twin silencers keep the exhaust free without entailing the necessity of a single expensive silencer.

Compression was not too high; cheap fuel took the machine along nicely and the engine did not bark so camming must have actually been pretty mild, so I put the problem down to a light flywheel. I could see the rim spinning around under its cast alloy cover on the drive side but not enough to gauge its size. I suspect, although have no way of knowing, that 'Guzzi lost some of its weight during the old singles face lift in an attempt to liven it up.

'And this they have certainly achieved, although not so much from the engine as from the cycle parts. After riding around on 750s, the 500 felt like a 250, especially as it was so low. Roadholding was just about perfect. Long bends or line chopping, they all came up roses. And that really is all I can say about roadholding. There it was — good — and what else is there? Town traffic trickling was moped-like and 80 mph cornering like 65 on most other machines. Any mediocre rider could give a man on a far higher powered machine a run for his efforts as long as the roads were bad — or good depending on your point of view.

Suspension was firm, like most good Italian machinery, although not uncomfortably so. Softish springs but firm damping, just the way it should be.

Braking of course, in Guzzi tradition, was enough for an engine of twice the power. Twin leading shoes in an eight-inch drum did everything (well, almost) a man could want. No fierce grabbing and no unwanted servo-action; just pressure and action in well balanced quantities. I don't know about waterproofing because it was around 90 deg F when I rode this. The electrics are in the same boat too, although the horn did work, but only in the manner of horns that hate to be worked, a sort of ill-tempered yap, like a squeezed Pickaknees.

Unfortunately, so much of this impression has to be guessed at because so little info' about the machine is available, although undoubtedly Don Barrat or Rivetts would get you one should you want it, and believe me, despite my criticisms of its bottom end performance, which in fact amounts to little more than nominal observations, it would make an extremely practical and thoroughly satisfying machine to own, let alone ride. Much more so than 75 per cent of the available machinery right now. I mean, it's top end behaviour is beyond reproof, more than equal to our own old singles. About the same as a T100 or a 441 Gold Star (can't bring myself to call that thing a

"Goldie"), y'know, good, great in fact.

It was a single after all, so I didn't change gear without the clutch more than once, and that carefully, 'cause it really ain't the way to keep new friends especially if they have MV fours they say you can ride if you come back on the Falcon with an open heart. But all in all, the complete transmission performance was like the rest of the bike — beyond criticism. It did **exactly** what it was supposed to and no bad habits.

What I did like was the electric starter. On a socking great single? You've just got to believe it. Exhaust valve lifter use or not, the half horse motor knocked the engine over compression as though it was not there. Better for the battery if the valve lifter was squeezed for the first compression stroke only, and then dropped once all was spinning freely obviously, but not essential.

Oh yes, almost forgot, bless someone's cotton socks, but the gear change was a proper one. None of your rotten, rotten, rotten Japanese/American rubbish of left side up for up (like cars my foot) but good old grand prix proven right side down for up. I love 'Guzzi just for that. I've got used to the present norm and have almost accepted it, but the Falcon let me know just how abominable it really is. Do Triumph realise what a filthy tradition they started when they started assembling gear boxes upside down and frightened Honda into doing the same thing?

There is not much more to say. Fuel consumption must be pretty good for it, always is on lightweight big singles, only I don't actually know. Finish was typically Italian. Great alloy castings, but only fair on paint and chrome.

For a guy who wanted more than riding thrills the Falcon has it, although I suspect that import duty, or whatever it is now we are all snarled up in VAT would not make the bike the bargain that it is in its own country. A great pity, because it's a machine that, at the right price, and I don't mean a ludicrously cheap one, could catch on within the group of riders who enjoy their motorcycles as machines.

The lack of a specification list is due to the lack of available specification, but this is what it basically comprises.

Type, all alloy flat single.

Capacity, 499 cc.

Bore and stroke, 88 x 82 mm.

Valve operation, pushrod.

Crankshaft, built up with all roller bearings and outrigger flywheel.

Carburettor, 29, Dell Orto breathing through paper filter.

Power output (DIN) 26.4 bhp at 4,800 rpm.

Electrics. 12v alternator, coil ignition, electric starter.

Transmission, gears, 13.23:1 7.42, 5.51, 4.12. All chain, wet clutch.

Frame, all welded full loop.

Tanks, fuel, approx 3 gallons, oil, 4½ pints in wet sump.

Wheels, front, 3.25 x 18in steel rim and 8in 2ls brake. Rear, 3.50, 8 in sls brake. m

May 20, 1973, will be remembered for many years throughout the motorcycling world as Black Sunday. As we were going to press with this issue we learned of the deaths at Monza of two great motorcycle personalities, Renzo Pasolini and Jarno Saarinen. May we take this opportunity to express our deepest sympathies and condolences to their families. They were both great sportsmen and will not easily be replaced.

WIXOM BROTHERS MOTO GUZZI 850 ELDORADO

A Transcontinental Explorer With A Touch Of Elegance Thrown In For Cruising The Boulevard

Cycle World Road Test

■ WHEN IT COMES to touring, lots of bikes will do, but few can match the stamina or comfort of a Wixom Brothers equipped Moto Guzzi 850.

The Eldorado, you see, is one of motorcycling's heavies. It's a bike for crossing continents, not states. It's a bike that can carry 200 lb. of gear for camping and a passenger at the same time. And, it's a bike that can eat up 300-mile sections of expressway and leave the rider free from fatigue.

The Guzzi is a sound machine all right, but appearance wise, it takes some getting used to. Fenders are flared and effectively enclose the 4.00-18-in. Pirelli tires fitted at both ends. And a massive front end and bulbous fuel tank, and you have a machine that closely resembles a Harley 74.

It's the kind of styling you either love, or hate. There is no in-between. If you can dig it, our test Guzzi will really interest you as it's a "dresser" in the full sense of the word.

Up front, there is a Wixom Brothers Ranger Fairing, complete with optional turn signals and pinstriping. Retail price of the fairing is $69.50, less turn signals which run $7.50 more. Pinstriping is an additional $5.

Complimenting this at the rear is a centrally mounted travelcase and twin saddlebags. Like the fairing, the luggage components are both color coordinated and pinstriped. The travelcase retails for $69.50, less pinstriping and running lights, which run an additional $5 each. Saddlebags are slightly more at $104.50 and again pinstriping is an extra, at $7.50.

The full dress look is expensive, but it's worth it if you spend a lot of time on the road. And, the majority of Wixom components are pretty trick. Take the saddlebags, for example. They are mounted on sturdy brackets, just as you'd expect, but Dzus fasteners allow detachment from the brackets in a matter of seconds. Consequently, you can use the bags like suitcases and carry them right into your motel room or whatever.

The travelcase is the same way. Undo four Dzus fasteners and take it with you. If, on the other hand, you prefer to leave the bags and travelcase mounted, they lock. As a final touch, the forward portion of the travelcase is padded and doubles as a back rest for the passenger!

Climb aboard the full dress Eldorado and the first thing you notice is the superbly contoured, amply padded seat. It's comfortable to begin with and remains comfortable, no matter how long you spend in the saddle.

The fuel tank seems enormous, but its contour blends in with the front of the seat and it in no way interferes with the rider. The portion of the tank which contacts the rider is chromed to preserve the finish on the machine.

Footpegs are properly placed and are comfortable for riders of all sizes. An immense brake pedal is located on the right. A rocker-type shift lever is on the left.

Grasp the bars. They are fairly narrow and high, forcing an upright riding position. With the Ranger Fairing, this isn't objectionable since it isn't necessary to lean forward into the wind. Without a fairing, most riders would opt for somewhat lower bars.

Handgrips are thick and soft for optimum comfort, but handlebar mounted control switches are awkward, if not impossible to use. On the left there is a rocker-type switch for high and low beam and a similar switch to activate the lights. In order to operate the high/low beam switch, the rider must remove his hand from the grip. This is not only annoying, but also dangerous on mountain roads.

On the right, a similar rocker-type switch controls the turn signals. Again, the rider must remove his hand from the grip (the throttle in this case) to operate the switch. Actually, it's easier to use the left hand for turn signals, and this is in no way satisfactory.

Just forward of the bars, there is a centrally located dashboard which contains a covered ignition switch and four idiot lights. For the first time, the lights, which indicate neutral and warn against overheating, low oil pressure, or faulty generator, are coded with a letter to indicate their function. On previous models, owners either had to memorize the position of the lights, or guess the ailment if one came on.

In typical motorcycle fashion, a tachometer is located to the left and a speedometer to the right of this panel. The tachometer is interesting in that it is electric and not mechanically driven by a cable, as is usually the case. Unfortunately, the needle wanders quite a bit, especially at low rpm or when executing quick shifts. Like the tach, the speedometer is not overly accurate. Also, the odometer is not resettable, so the rider must make a mental note in order to properly judge refueling intervals.

Lean the Eldorado upright, retract the sidestand, and plant both feet firmly on the ground. Even short riders can reach the ground on this one. Ground clearance is absolutely minimal. Flip on the ignition switch, advance the handlebar mounted choke slightly, and press the starter button. As soon as you notice the starter groan, the V-Twin fires and settles down to a somewhat erratic idle.

As with previous Guzzis, no kick starter is fitted and quite frankly, none is needed. The battery is an overly large, 32 amp/hour unit and it drives an 0.7 bhp starter motor. Any time an 850cc Twin can't be started with this setup, it's time to start looking for other problems!

Almost immediately after starting, the choke can be

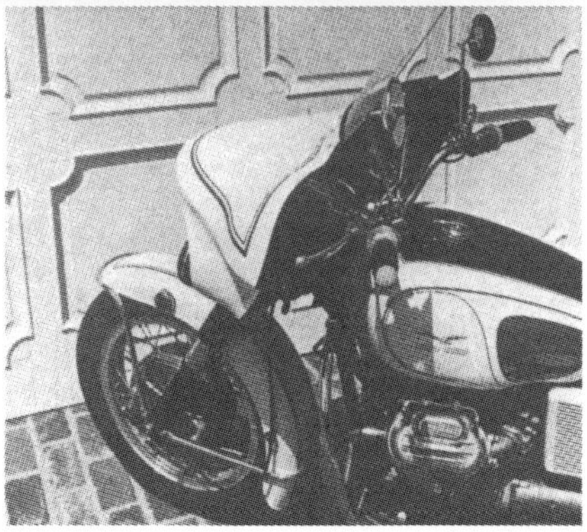

retarded and the idle will smooth slightly. Pull in the clutch and ease the Guzzi into gear. The shift pattern is just the opposite of most bikes. It's up for low and down for the remainder of the five speeds.

As soon as the Guzzi gets underway, the cumbersome feeling disappears. The Eldorado is actually quite agile in town; surprisingly so on turnpikes or the open road. No doubt this is why several police officers favor Guzzis for traffic control duty.

Enter a bumpy turn, though, and problems arise. The suspension can handle it, but the massive chassis is too low to the ground. Consequently, is scrapes. It scrapes in left turns, right turns, and sometimes when entering driveways straight on. The funny thing about it is that Moto Guzzi realizes the problem, and has built steel skids into the forward portion of the crankcase to enable it to slide over obstacles and protect it from damage.

Fortunately, at high speeds, the problem is not as acute. It just isn't necessary to lean the bike over far enough to cause a problem. To prove it to ourselves, CYCLE WORLD staffers spent several hours on mountain roads. Handling is actually superb for a bike of this size, allowing any combination of sweeping or hairpin turns to be taken at a healthy rate of speed with confidence. Sure, the Guzzi can be grounded, and hopefully this will be remedied for an additional safety margin, but you have to exceed the limits of the law to get into any real trouble.

Mountain roads didn't shake our confidence in the handling, but they did draw attention to the brakes. The rear unit is overly sensitive, probably due to the inordinate amount of leverage made possible by the design of the rear brake lever. Contrasting this is the front twin-shoe drum which fades after a couple of panic stops and requires a great deal of lever pressure any time it's needed. This is unusual in that previous Guzzis we've had for test have proven superb stoppers with similar brakes.

It may seem strange to harp on the brakes of a touring machine, but one must realize that the Guzzi is capable of cruising along at 100 mph. Acceleration is brisk as well. Just roll the throttle on. Any gear will suffice. Shifting isn't really necessary. There's that much usable torque and power on tap.

When shifting does become necessary, however, it isn't the pleasurable task found on most machines. The gearbox clunks slightly, although nowhere near as severely as on previous Guzzis, and considerable pressure on the pedal is required for gear changes. This clunk when shifting won't hurt the transmission, but it is decidedly annoying in view of the engine's silent running.

More annoying than the clunk is the fact that the neutral light comes on occasionally when the bike is still in gear. Let out the clutch rapidly and you'll find out just how awkward an Eldorado can be when it's tipping over.

While we're on the subject of handling and riding, a few words about the Wixom accessories are in order. The Ranger Fairing is simply fantastic. It protects the rider's entire upper body, hands included, from 90 percent of the wind, regardless of the speed involved. And, handling isn't affected at all until the 100 mph mark is passed. Above 100, the bike oscillates slightly, but not enough to cause alarm. Without the fairing, the oscillation above 100 never appears. In fact, the Guzzi Eldorado is one of the most stable motorcycles on the market today.

The saddlebags and travelpack are equally impressive. When empty, they are not noticeable at all. With a moderate load, this remains true, although drag racing starts will produce wheelies if anything in excess of 80 lb. is carried in the Wixom components.

In spite of the Eldorado's full dress styling and capable handling, the machine's power train remains one of its more interesting aspects. Highly automotive in character, the engine is a 90-degree V-Twin mounted transversely in the frame, much like a BMW.

A massive forged steel crankshaft rides in automotive-type plain main bearings and carries the connecting rods, one beside

WIXOM BROTHERS MOTO GUZZI 850 ELDORADO

SPECIFICATIONS

List price	$2189
Suspension, front	telescopic fork
Suspension, rear	swinging arm
Tire, front	4.00-18
Tire, rear	4.00-18
Brake, front, diameter x width, in.	8.7 x 1.7
Brake, rear, diameter x width, in.	8.7 x 1.7
Total brake swept area, sq. in.	92
Brake loading, lb./sq. in. (160-lb. rider)	8.2
Engine, type	four-stroke ohv V-Twin
Bore x stroke, in., mm	3.27 x 3.07, 83 x 78
Piston displacement, cu. in., cc	51.7, 844
Compression ratio	9.2:1
Claimed bhp @ rpm	64.5 @ 6500
Claimed torque @ rpm, lb.-ft.	N.A.
Carburetion	(2) 29mm Dellorto VHB
Ignition	coil and battery
Oil system	gear pump, wet sump
Oil capacity, pt.	6.5
Fuel capacity, U.S. gal.	5.8
Recommended fuel	premium
Starting system	electric
Lighting system	12V generator
Air filtration	dry treated paper
Clutch	twin-disc, dry
Primary drive	gear
Final drive	shaft
Gear ratios, overall:1	
5th	4.28
4th	4.96
3rd	5.99
2nd	7.93
1st	11.42
Wheelbase, in.	57.9
Seat height, in.	32.1
Seat width, in.	12.5
Handlebar width, in.	31.0
Footpeg height, in.	11.9
Ground clearance, in.	4.9
Curb weight (w/half-tank fuel), lb.	581
Weight bias, front/rear, percent	45/55
Test weight (fuel and rider), lb.	721
Mileage at completion of test	2001

TEST CONDITIONS

Air temperature, degrees F	75
Humidity, percent	58
Barometric pressure, in. hg.	30.25
Altitude above mean sea level, ft.	383
Wind velocity, mph	2-5
Strip alignment, relative wind:	

PERFORMANCE

To speed (actual @5600 rpm), mph	98
Computed top speed in gears (@6500 rpm), mph:	
5th	114
4th	98
3rd	81
2nd	61
1st	42
Mph/1000 rpm, top gear	17.5
Engine revolutions/mile, top gear	3432
Piston speed (@6500 rpm), ft./min.	2975
Lb./hp (160-lb. rider)	11.5
Fuel consumption, mpg	48
Speedometer error:	
50 mph indicated, actually	44
60 mph indicated, actually	54
70 mph indicated, actually	64
Braking distance:	
from 30 mph, ft.	34
from 60 mph, ft.	158
Acceleration, zero to:	
30 mph, sec.	3.0
40 mph, sec.	3.9
50 mph, sec.	5.0
60 mph, sec.	6.1
70 mph, sec.	8.5
80 mph, sec.	10.9
90 mph, sec.	15.8
Standing one-eighth mile, sec.	9.86
terminal speed, mph	75.44
Standing one-quarter mile, sec.	14.49
terminal speed, mph	86.50

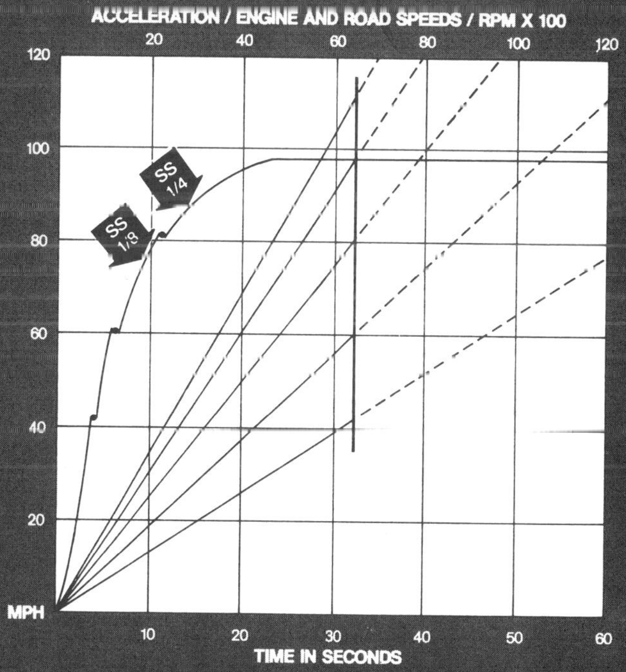

MOTO GUZZI

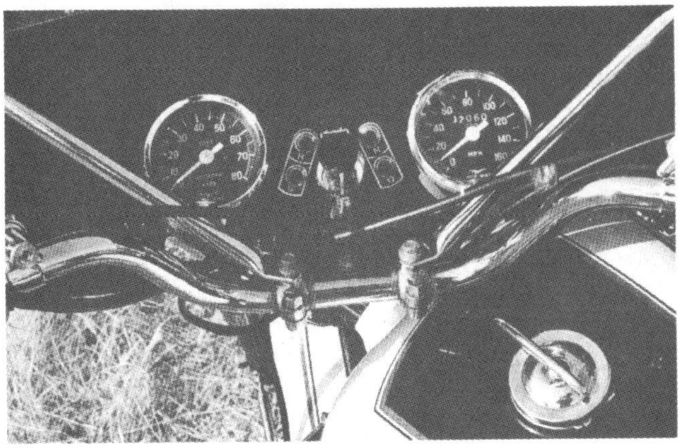

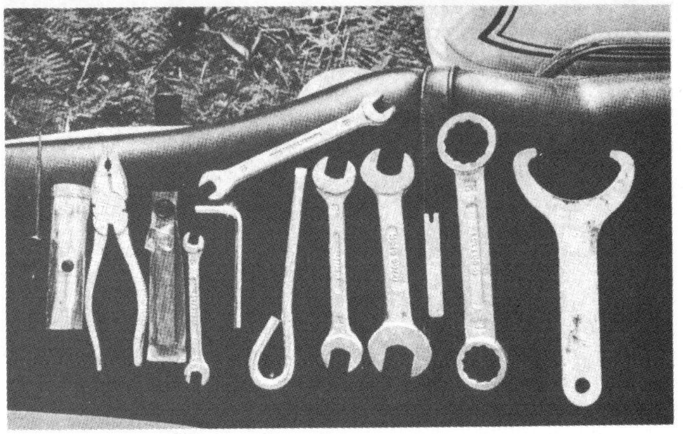

the other on the same crankpin. The 90-degree layout allows balancing to eliminate primary vibration, although it does allow a slight amount of rocking couple which is not present in the BMW's 180-degree layout.

Plain bearing inserts are used in the connecting rod big ends and bushings carry the piston pins. The high oil pressure required for such a setup (54 to 60 psi) is provided by a two gear oil pump. The pistons have two compression rings and two oil control rings. The final oil control ring is unusual in that it is located below the piston pin. The pistons are set up to run at a very small clearance in the chrome plated cylinder bores.

Light alloy cylinder heads feature hemispherical combustion chambers and special cast iron valve seats. Two 29mm Dellorto concentric carburetors with square slides provide the fuel/air mixture and these are tucked in toward the centerline of the machine to prevent the rider's legs from hitting them.

The high mounted camshaft is gear driven by the crankshaft and operates the valves through tappets and short pushrods. Even though the Guzzi is not a high rpm engine, a short valve train makes it possible to use lighter valve springs with resultant smaller amount of wear on the camshaft, rocker arms and valve guides.

The area between the cylinders is filled with a heavy-duty automotive type Marelli generator. It's rated at a healthy 300 watts and is driven from the crankshaft by a rubber V-belt. The lower pulley has notches on it to correspond to proper ignition timing at different engine speeds, making it easy to check the timing with a strobe light.

A twin-plate dry automotive-type clutch separates the engine from the five-speed gearbox which is a moderately close-ratio unit. Heavy-duty gears are changed by a selector drum, but silent gear changes require allowing the engine and road speeds to synchronize because the straight-cut, non-synchronized gears are spinning at engine speed.

A shaft drive is totally enclosed, silent, and requires a minimum of maintenance. It is slightly less efficient than a chain because of the bevel gears, but the rear wheel stays clean and the rider doesn't have to worry about getting oil all over his back. Further, there are no periodic adjustments, cleaning, or lubricating drills. Aside from changing the rear end oil every 6000 miles, the secondary drive train can be forgotten.

Because the Guzzi V-Twin is simple, periodic maintenance is easy to perform. In fact, any knowledgeable home mechanic should be able to service his machine with the aid of the well written and illustrated rider's handbook. Large filler and drain plugs simplify oil changes, and all items requiring adjustment are within easy reach.

CYCLE WORLD has reports of riders going in excess of 70,000 miles on their Guzzis with little more maintenance than changing the oil, periodic tune-ups, and tire replacements. Like we said, the Moto Guzzi 850 Eldorado is a machine for crossing continents. As miles pile up, most other machines will have trouble maintaining the pace.

GUZZI'S SUPREME SPORTSTER

Guzzi's finest bike — a masterpiece even more rare than the fabled Agusta — is known to only a few riders in the world. It is, in many ways the perfect motorcycle. Two a week are handbuilt at Guzzi's Milan factory. TWO WHEELS has ridden one — the only one in Australia. And it's likely to be the only one to ever come to Australia. It's $2700 worth of machinery; big, beautifully functional, exclusive. Here's what it's like to ride . . .

Right: The Guzzi is interesting dimensionally — and a lot smaller than the bulky appearance resulting from the unusual motor. Bike weighs only 468 lb.

Below: Motor is rated at 72 bhp at 7000 rpm, runs at 9.8:1 compression.

Above: Mufflers are removable-baffle Silentium. Shaft drive also forms member of swing arm. Both back and front footrests are on the same bracket — so they can be easily removed for rearsets.

WHILE most of the Moto Guzzi factory concentrates on the production of big lusty touring 850 twins, a handful of the most experienced staff hand build the most luxurious 750 sportsters in the world.

They leave the factory at the rate of only two a week and receive exhaustive track testing before being exported.

There is only one in Australia and to get it here, distributor Ron Angel had to visit the factory in Milan, Italy and shoehorn it out of them.

After uncrating in his workshop he just sat the bike on the showroom floor and left it there, not wanting to sell and waiting for the day when he would have the time to ride the bike himself . . .

In many ways sitting on the V750 Sports is the only way to realise just how unusual this bike is. The 29 in. seat height is amazingly low for a 750, and with the *fully* adjustable handlebars placed in the down position, the whole machine feels like one of the 250s we test later in this issue.

And now to start it. There is no need to turn on the petrol — an electric petrol tap does that automatically! All the rider has to do is turn the key which unlocks the steering and switches on the ignition. The "systems go" signal is a click from beneath the petrol tank.

No kickstart lever is fitted, but there's two ways of electric starting the bike. One is by turning the ignition key further (like a car) and the other is by pressing a thumb button under the twistgrip. Very neat.

Like all transverse-mounted motors, blipping the throttle of the Guzzi at tickover produces a sideways torque reaction. Despite the light weight of the bike, the movement is no more than that of a 750 BMW or the touring version of this Guzzi.

A combination of a two-plate clutch and a well-thought-out gearbox rids this Guzzi of our usual grumbles of transverse mounted twins. This V750 Sport engages first *silently* and can be shifted *silently*.

The big V-twin begins to sort itself out at 3500 and by 4000 all is smooth, leaving a 3000 rev band to maximum power at 7000. The gears are extremely close and on fast, winding roads, where the handling comes into its own this V750 Sport is a very fast machine.

And although the makers have fitted excellent suspension, an exemplary frame, a low centre of gravity and 72 bhp powerplant, the bike is still very much a tourer. The present owner bought it to replace his BMW because he was tired of the limitations of BMW performance and gearshifting. This Guzzi has all the same qualities that BM riders love — smoothness, shaft drive and dependability, yet it has a lot lot more thrown in as well.

The handlebars can either be extended to the fully-up around-town-comfort position or dropped six inches and swept back like clip-ons. *(Continued on page164)*

GUZZI'S SUPREME SPORTSTER

In this way the rider can totally alter the feel and character of the bike to suit the riding he wants to do.

Whether that riding be flat out at 120 or smoothing along at 60 this Guzzi will do it — and for a long, long time. The bike is the result of "no costs spared" individualistic design that has been handbuilt by master craftsmen. We spent a whole afternoon starting at the front and working backwards just looking at the attention to detail.

SPECIFICATIONS

MAKE	**MOTO GUZZI**
MODEL	**V7 Sport**
PRICE	**$2695**

ENGINE:
748 cc 90 deg transverse twin. Gear drive from crank to single central cam to overhead valves set in hemispherical heads. All-aluminium construction with each chamber vented to atmosphere. Roller main bearings and shell big ends, 9.8:1 compression with 30 mm Del Orto carbs producing 72 bhp at 7000 rpm.

TRANSMISSION:
Twin-plate dry clutch mounted on crankshaft end. Five speed constant mesh gearbox. Overall ratios: 4.05:1, 4.69:1, 5.65:1, 7.49:1 and 10.8:1.

FRAME:
Double loop cradle, fully triangulated with the bottom section detachable to allow engine to be removed.

SUSPENSION:
Front: Guzzi forks (two-way damped).
Rear: Koni units (three-way adjustments).

BRAKES:
Front: 9 in. double sided four-leading shoes.
Rear: 9 in. twin leading shoe.
Wheels: 18 in. alloy rimmed (interchangeable).

CAPACITIES:

Fuel tank	4½ gallons
Oil tank	.6 pints
Transmission	1½ pints

DIMENSIONS:

Weight	468 lb
Wheelbase	59 in.
Overall length	86 in.
Overall height	40 in.
Overall width	27 in.
Seat height	29 in.
Ground clearance	7 in.

PERFORMANCE

ACCELERATION:

0 to 60	5.8 sec
Standing ¼ mile	13.1 sec

SPEED:

Maximum	126 mph
Comfortable cruising	50 to 110 mph

FUEL CONSUMPTION:

High speed	40 mpg
Cruising	51 mpg

BRAKING:

From 30 mph	25 ft

HOW LONG DOES IT TAKE

Change plugs	1 min
Remove rear wheel	4 min
Adjust contact points	5 min
Check transmission oil	2 min
Check tappets	20 min
Adjust brakes	2 min
Top up battery	1 min

HOW DO THE FINER POINTS SCORE (out of 10)

Instrumentation	10
Seat comfort	9
Tool kit	8
Quality of finish	10
Vibration	9
Lighting	10
Positioning of controls	10
Spares availability	2

The Moto Guzzi V750 Sport is the only bike we have ever tested that has no faults. We can honestly say that designers have achieved their object in building a perfect tourer/sportster that combines durability with speed and comfort.

Though the engine is basically the same as the touring Guzzi each component has been lightened and polished before assembly. The bore is also 0.5 mm smaller to bring the capacity down to 750 cc.

As well as the stage of tune being high, there are many other interesting features to the motor. The lubrication system works at twice the normal rate and each chamber, (including cams and rocker boxes) is individually vented through a common collector box that extracts any oil and returns it to the sump. To ensure minimum weight, a special Bosch alternator and starter motor have been fitted which, although being small, are very efficient, and yes, you've guessed it, very expensive. No distributor (as on the tourer) is used. Instead there are two sets of points and two coils. As can be expected the whole motor is put together with Allen screws.

No air filter is fitted as standard but the small air box between the carbs has provision for one and the owner has now fitted an oiled-foam element.

The Guzzi's gearshift shaft enters the box at the rear of the left side and the shift lever is on the right side, connected in a linkage system. Unlike normal practice where a pivot point is a drilled fork and cotter pin, the Guzzi has an excellent system using SKF sealed swivels. These will last a long time before any slop develops and can be replaced easily and quickly to regain perfect linkage control.

A look around the engine/gearbox unit shows the heavy-tubed, well-triangulated frame to have been built around the aluminium structure. The two bottom tubes can be detached from the rest of the frame, allowing the complete engine and transmission to be dropped away in less than half an hour.

The rest of the bike is certainly just as fascinating. The electric petrol tap is complemented with a manual one for reserve and also if the electricals fail. The 4½-gallon tank is deceptively small in appearance can be removed in a couple of minutes and is well styled to fit in with the rider's knees and seat.

Releasing a catch at the left side front of the dual seat allows it to swivel back. This triggers a courtesy light under the saddle and gives access to the huge battery and comprehensive CEV fusebox which has provision for a large number of take-off points should the owner choose to fit a number of electrical accessories.

Loosening two small thumb screws at the back of the seat permits the whole unit (back mudguard included) to swing up out of the way so that the back wheel can be rolled clear.

Just beneath the saddle and either side of the battery are two tool boxes — that lock. The standard kit is good, and comes wrapped up in a long-lasting heavy fabric bag. Also at the back end is a well-sealed rear stop light, removable baffle (Silentium) silencers, twin grab rails, a twin leading shoe 9 in. rear brake (wheels are interchangeable), fully adjustable Koni units, and a very clean-looking final drive assembly.

Rider pegs, foot controls and handlebar rubbers are all of a matching pattern with the Guzzi name stamped on them. Neat.

An awful lot happens around the steering head. Going through it is the damper control which varies the leverage length of the arm that actuates the telescopic fluid damper — the best steering control we have ever seen. Just behind is the master switch which turns on the ignition, actuates the starter motor, and locks the steering when the key is removed. Neat again!

Both triple clamps are huge alloy castings finished in crackle black and like the rest of the bike, they have chrome-plated Allen screws. The instruments are rubber-mounted and as well as the Veglia dials being accurate they are well illuminated.

The handlebar clamps slide up and down the top of the fork legs to enable them to be set literally to any position the rider chooses. Both handlebar levers are beautifully finished alloy and have quick and effective adjusters.

Though the front forks are of a completely new design they are modelled on the Ceriani pattern and every bit as good. Their 5½ in. movement is progressively sprung and well damped with absolutely NO oil leakage. Even the oil drain plugs are plated Allen screws!

Like most Italian machinery the mudguards are stainless steel and the rims are Borriani alloy, but unlike the others the guards are hand polished to a chrome-like finish and the rims are of a stronger section.

The front brake is a double-sided, four leading shoe assembly, which, with hard linings and ample cooling gives good feel and fade-free stopping power.

Briefly — and we mean briefly — that is the Moto Guzzi V750 Sport. In many respects it could be the *perfect* motorcycle.

Our thanks to the new owner, Geoff Read, for his co-operation in making our test possible. We understand the sale included an agreement to allow Ron Angel to have an occasional ride . . . *

MOTO GUZZI

If a Sherman tank was quiet, smooth and fast . . .

Type Usage	Displ.	Engine Type	Bore & Stroke	HP@RPM	Comp. Ratio	Carburetion	Ignition	Gears	Wheel-Base	Dry Weight	Aprox. Speed	Aprox. Price

V7 SPORT

Legions of purists tantalized by successful European racing efforts with slimmed-down Guzzi 750s have waited a long time for the V7 Sport, an off-the-floor cafe racer supreme. The bike offers all the rugged reliability originally designed into the touring and police workhorse, but wraps it in a more nimble, lighter package with superb styling and finish. A giant double-plate dry clutch and shaft drive are major drive-train advantages of the Sport over other bikes. Alloy rims, Michelin tires add to its class.

Street	748cc	4/S ohv V-Twin	82.5x70.2mm	70@7000	9.8:1	2,30mm, Dell'Orto	Bat. & Coil	5	57.8"	495 lbs.	115 mph	$2389

EL DORADO

In mid-1972 Moto Guzzi upped the displacement of their Ambassador 750 touring bike to 850cc and called the new machine El Dorado. It has 4½ more hp, an extra gearbox speed, beefier crankcases and a stronger rear hub with more oil capacity. The bigger size proved so successful on police bikes (available to individuals on special request) that Guzzi made 850ccs their exclusive touring displacement for '73. Huge plain main bearings and a one-piece forged crank helps give the 850 its world-renowned reliability.

Street	836cc	4/S ohv V-Twin	83x78mm	65.5@6500	9.2:1	2,29mm, Dell'Orto	Bat. & Coil	5	59"	530 lbs.	110 mph	$2004

POLICE SPECIAL

Believe it or not you as a private individual can buy a police-equipped Moto Guzzi 850 from any Guzzi dealer. The special equipment makes the bike touring-ready without any further effort on your part and includes floorboards, lock-out sidestand, a dash panel, taller and wider handlebars, rear safety bars and your choice of fairings, windshields, saddlebags and extra lighting. Weight is about 70 pounds greater than a standard Eldorado and price will be about $100-200 more. Colors are black, white, two tone.

Street	836cc	4/S ohv V-Twin	83x78mm	65.5@6500	9.2:1	2,29mm, Dell'Orto	Bat. & Coil	5	59"	600 lbs.	110 mph	NA

"How's the Guzzi Sport grab ya?" Every day the inevitable question became more traumatic, for with each passing mile it grew increasingly impossible to answer in the normal context of conversation. Like the alcoholic who mistakenly downed a fifth of cod liver oil sitting next to his bottle of gin, you're afraid to sneeze; anything more than a smile and a pat on the tank is apt to trigger total response.

Help me sneak up on it this time. We'll start real cool; like "The styling's rather different, isn't it?" Yup, you might say that it was designed on purpose, that the stylist and the engineer were definitely on a first-name basis. It isn't accidental that the angle of the V-bottomed gas tank exactly conforms to the cant of the cylinders, or that the slant of the double-loop frame's vertical tubes closely corresponds to that of the rear shocks, giving this racehorse that speed-at-rest look. Those Italians! They broke their pick on this one. Styling is tight and fat; each component, each group of components complementing the other. To rope and tie the youthful American mag wheel faddists, flat-sided, deep-wall Borrani rims with maximum reflective surface were kept to minimal 18-inch diameter, then stuffed full with massive finned alloy drums that respond to the crook of a finger and tap of a toe. Discs? They'd only amount to overkill compared to this four-leading-shoe front drum; and aesthetically it would be tantamount to bolting on a couple of dinner plates to the bottom of each leg—ugh. And rather than run spindly contemporary braces to the far reaches of the fender, the tight-fat theme is seen again in the close-cropped Guzzi brackets; their oversize diameter permits them to be kept short and complementary rather than long and unsightly. Crowning frontal appearance is a set of extremely comfortable clip-on handlebars, adjustable for height and sweepback.

Guzzi's attention to detail is unparalleled. It is reflected in such styling tools as symmetry, finish and metal forming technique. The headlight brackets, rather than thin steel stampings, are sculptured angular castings that tuck immediately between the fork legs and disappear so as not to mar the vertical line of the forks and bars. Playing down their existence further is a dull black crackle finish identical to that of the fork triple clamps. Guzzi obviously detests the over-use of chrome, as reflected in the satin gloss of eternally tarnish-proof stainless steel fenders. And symmetry abounds, in the shift and brake pedals that are mirror copies of one another, the dual horns that speak with the authority of a European motorcar rather than a motorcycle, and the twin toolboxes that have the capacity to pack ample wrenches plus a couple of sandwiches under tumbler lock and key. Would you go out of your way to position the fork top pinch

For those who thought they couldn't afford a Ferrari.

Motorcyclist Test by Bob Greene

MOTO GUZZI V-7 SPORT: RIGHT ON!

bolts—socket head bolts, at that—inboard of the triple clamp rather than at the outside edge just for appearance sake? Guzzi did, and it is this multitude of minute touches, that only a connoisseur would catch, that provide the clue to Milan's in-depth concern for the ultimate affect. Obviously they are as serious about their obligation to you as yours to them: $2500.

Is the Sport Guzzi anything at all like its brother 850 El Dorado? Nothing at all. Aside from being 100 cee's smaller, to scootch under the 750 race track limitation, it has a more cammy cam and notched pistons to accommodate the lift, plus a higher compression ratio of 9.8:1. And the flywheels feel lighter than either the 850 or the old 750 Ambassador. Another difference is the use of an alternator rather than a generator. More noticeable, however, is the V-7 Sport's unique full-loop frame with removable bottom rails to facilitate major engine overhaul. Rather than the bolt-upright "witness chair" riding position of the Dorado, the Sport is a lush body-hugger; it embraces the human form in a most comfortable—almost loving, if that were possible—manner that creates the illusion of bod and bike having been built in the same factory, at the same time. As opposed to the bump-on-a-log riding position of the garden variety motorcycle, the Guzzi V-7 handler drops into a saddle barely 29½ inches off the deck, sitting down in behind the machine in such a manner that rider and mount become one, moving together rather than his waving around high in the air. Feet are back, knees slightly bent, arms somewhat down and closer together than usual, with the body in an attitude that most easily permits the rider to rise out of the saddle should it be necessary. To those who feel that such a riding position would be uncomfortable, hear this: Just six months on the mend from a broken back, I was able to comfortably tolerate only this chassis among all of the big-bore tourers for extended riding jaunts up to 100 miles or so. It's one of those try-it-you'll-like-it situations, I thought it would be too sporting for my blood, too, but it turned out to be just the reverse. Should you want to sit right up, it is only necessary to slide forward to the nose of the saddle.

Contributing to this old-shoe comfort is a firm but gentle saddle and a pair of three-way-adjustable Koni rear shocks that are standard appointments. When an Italian is able to subdue his national persuasion to the point of using Dutch suspension in the pursuit of perfection, that shows me class. Not only are the shocks of the finest quality and precise progression but the spring coils extend the full length of the shock body, these two factors combining to give a fabulously cushioned ride that defies bot-

Tightly grouped instrument cluster includes 8000 rpm tach and 160 mph speedo, but no trip mileage cluster. Shame! Headlight makes day out of night, but electrical hand switches are very difficult to finger. Fortunately they can be replaced rather easily with Yamaha 650 switches which fall easily to hand. Check that pop-top gas cap and trick key which activates ignition, forks, fuel valve, starter with one turn!

PHOTOGRAPHY BY ERIC HICKMAN

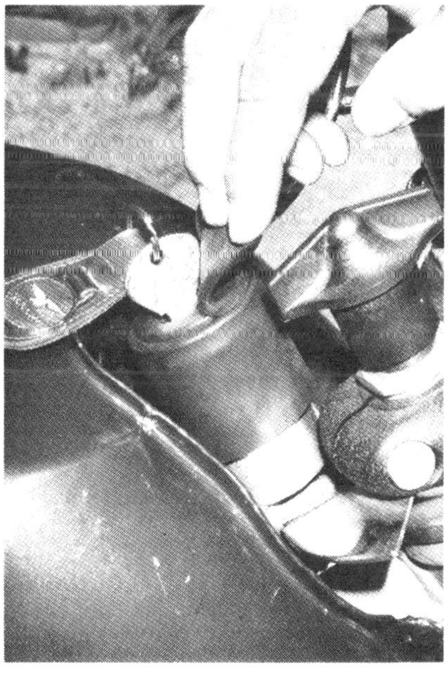

toming even with two up! If two big fatties jump up and drop flat into the saddle, there is seemingly no stop to the downward travel. The shocks can be seen to use up two or three inches of their four-inch travel, but the sensation of the rider is merely akin to doing a pratfall into a tub of glue. It appears the shocks' reserve is more than would ever be encountered in a street situation. Compared to the shocks on a nearby Kawasaki Four, for example, the following dimensions tell the story: The Konis have 1 7/16-inch body diameter, 4-inch stroke and 9-inch spring length; the Kawasaki shocks have 1 3/16 body diameter, less than 3-inch stroke and 7-inch spring length. The Konis' 1/4-inch larger diameter, 1-inch longer stroke and 2 inches more spring create a world of difference. Front fork travel, although slightly more Spartan, is also of the generous 4-inch category. Since the hydraulic shock damping units in these front forks are self-contained and sealed within the fork legs, only 1 3/4 ounces of oil per leg is required for lubricating purposes. The forks are unorthodox in this respect but do their work quite satisfactorily. The principle has merit since the damping oil, working inside a sealed shock body within the fork leg, is immune from normal contamination that passes by the slider seal of a conventional leg. Overall suspension and handling are superb, putting to rest that old crutch about a high-speed road machine having to ride like a dump truck to do its thing. In fact, the Sport is able to hang in the rough corners specifically because its wheels always have hold of the road. And there's nothing to hang up in the most violent turn; you're miles off the deck even in the raciest attitude through the swervery.

Handling is spot-on straight ahead or laid over, with never a hint of oscillation from road irregularity in a bank. In a cornering attitude the machine remains neutral, neither pulling down nor raising up. If you prefer, there is a steering damper, a hydraulic unit that can be turned on or off by a swing-away front pivot point that repositions the nose of the hydraulic tube through a quarter-turn of the control knob atop the steering post.

Should it be necessary to reign-in the Sport, there's a lot in your favor; 8.6 x 1-inch four leading shoe twin drums front and dual leading shoes of the same dimensions rear. Since you can get rubber with either brake, discs are not needed. They've got more than you've got. Brake action is very easy, very progressive and very assuring. The hubs reflect Guzzi's racing background, being finned and vented around their periphery rather than through the backing plates, thereby taking advantage of centrifugal force to purge the brake cavity of dust or water. And it is one of the extremely few modern motorcycles I can recall that beds the spoke heads in the hub proper, permitting an ideal end-to-end pull, rather than hooking them into a raised flange. The straight-pull spoke, of course, has the advantage of being immune to bending and loosening, retaining their tension indefinitely. Guzzi cares.

About that cross-threaded engine: Transverse positioning of the chrome-bore cylinders obviously provides maximum exposure to air flow, plus facilitating hookup to the shaft drive since crankshaft and driveshaft both run fore and aft. Saw the front two cylinders off your family V8 and you've just about got it; a pushrod overhead valve setup with wet sump and geared oil pump. Rods and crank ride on plain bearings, automotive style, with power delivery taken off a rear-mounted flywheel and two-plate clutch driving the gearbox mainshaft. Power from the mainshaft is passed to the adjacent layshaft and out through a universal joint to the driveshaft and then on to the rear wheel ring and pinion. Of 748.4 cubic centimeters (45.66 cubic inches), the 82.5 x 70mm bore-stroke twin has its cylinders cocked at 90° with both rods swinging on a common throw. Power is rated 72 hp at max revs—7000 rpm. Electrics are German Bosch, using alternator, battery and contact breaker points with automatic advance.

Because of the healthy battle that is now developing between the multis and the twins, let's talk about the transverse V for a minute. Although the Sport has a noticeable side-bumping action at idle, this shake disappears once the bike moves away from rest. If you're a mirror watcher, you'll rate the V-7 with or above the multis at cruise, for the reflected image of cars from behind is crystal clear. Vibration at cruising speeds is nil. And unlike any other, this is the only motorcycle in the world that doesn't have a chain—nowhere. The only chain you'll find on the Moto Guzzi V-7 is the key chain! And since at least three-quarters of the enthusiast's maintenance time is spent adjusting the final drive chain, the Guzzi owner has a lot of free time on his hands—clean hands, at that. The old-time chain-drive biker actually feels a tinge of guilt about the lack of maintenance—almost frustrated—until it finally sinks in that about all there is to do with the V-7 is wipe off the tank and top-up the battery. Although the cylinders are equally exposed to the air as a 180° opposed twin, cylinders and carburetion are completely out of the way of the rider's legs and feet, unless, on a cold night he opts to slide forward on the saddle and snuggle up against the rocker boxes to warm his knees—it works great. Engine and exhaust are exceptionally quiet, the latter just a whisper; much more so than any of the multis, and, therefore, less tiring over a day's ride. With the exception of a very slight mist from a

small vent hole in the gearbox breather cap—which should be vented to the adjacent rocker box breather chamber or down to near ground level—the engine and transmission case remained absolutely spotless during our 700 miles together.

But unlike the fours and triples, and several of the Guzzi's fellow 750cc classmen, it is no shotgun. Smooth, pleasant and—for me—exhilarating it is, but a Sportster it ain't. You get your kicks on the Guzzi from the precise, taut, rare manner in which it responds to your every whim; the positive, hot-knife-through-butter way it shifts; the soft, silky smoothness; and the intimate bike-rider relationship that has no counterpart. Performance is ample but not in the dragster category. The sheer brute torque, the hard-hammering whack that is absent in the lower rev ranges is compensated by chassis agility in tortuous territory. Although the engine pulls well enough at 2500 rpm in high gear on the level, it starts to make a muscle at around 4000, and in the mountains it is advisable to have about 5000 rpm in hand to maintain "charging speed." Actually the bike is overgeared; it will run faster in fourth than it will in fifth—uphill *and* down.

Although the factory claims a top speed of 129 mph, the fastest I ever saw, on a downhill run in fourth, was 110. Inclined to believe the factory in

Below left—*Change a tire? Simple with Guzzi's flip-up fender. And no chain adjustment to fuss with—ever. Full-length springs on giant Koni shocks provide a ride that has to be experienced to be believed! Like riding on a cloud.*

Below—*Exciting and symmetrical from any angle, the Guzzi V-7 was obviously styled by an enthusiast. The generous saddle is extremely comfortable.*

Right—*Massive four-leading shoe brake, with seldom seen straight-pull spokes that never loosen. It's really a super setup.*

Below right—*Huge battery, fuse center, and handy automatic trouble light.*

this instance, I can only attribute the difference to one thing: the fitting of an accessory styrafoam air filter sock between the two carburetors on V-7's sold by western states distributor ZDS Motors. The carbs are factory-equipped with high-performance velocity stacks which must be removed to accommodate the interconnecting oil-wetted air cleaner. ZDS's concern for longevity could be throwing a wet blanket on top-end performance by slightly restricting breathing to the extent of making the engine only appear overgeared.

But if it is truly overgeared, there is one fringe benefit: the lazy, almost dream-like, engine speed on the highway has a beneficial effect on not only the rider's ease of mind but engine accessory and driveline component life as well. Carburetor jetting seemed right on, with never a misfire or indication of heat. On rare occasions the machine would indicate 105 in fifth gear, and on one long, downhill run we came up with a 109 at 5900 in high, but during the specific period assigned to speed runs on level ground, the following figures were logged and serve as official for this test. All readings were made from the Guzzi's own tach and speedometer within about a one-hour period at the 2000-foot level in an 80-degree desert environment. These relative speeds and revs should provide a pretty close indication of overall performance.

MPH @ RPM in Fifth Gear		Max Speed @ RPM in Gears	
60 mph	3200	50 @ 7000	Low
70	3800	80 @ 7000	Second
80	4300	100 @ 7000	Third
90	4900	110 @ 6700	Fourth
100	5300	100 @ 5000	Fifth

With or without the velocity stacks, the V-7 cruises effortlessly at 80-90 mph; the engine runs half-speed at legal highway limits! That's probably close to 800 rpm less than the Kawasaki Four, when the Kaw's speedometer error is taken into consideration. Speaking well for the motorcycle's overall performance in a twisting, climbing situation, the Sport, even with its carburetor muffler in place, spirited this cautious, broken old rascal over his favorite 35-mile mountain pass in 33 minutes by stopwatch, at an average 63.64 mph—my best time to date. Not fast, the figure is only interesting because it is relative to previous performances on many, many brands of various displacement motorcycles over a long period of years by the same person, at a safe and sane pace. Seldom has that time bettered 35 minutes. Let's say the Sport's no slouch.

As mentioned at the outset, the Moto Guzzi V-7 Sport has such a bounty of practical, unique features that one hardly knows where to begin when placing them in their proper order of importance. Although the Sport's "Magic Key" happens to fall late in this report, to me this ingenious concept proved utterly fascinating. The ignition key automatically, and with a single movement, unlocks the steering lock, turns on the ignition, opens the electromagnetic fuel petcock and activates the electric starter (no kickstater is fitted). Thus three time-consuming manual operations necessary to start-up any other motorcycle are completely eliminated. Similarly, upon parking the machine, the rider merely removes the key and walks away; there is nothing else to do. The forks automatically lock, the gas automatically shuts off. A conventional handlebar-mounted starter button is fitted, however, which can be used in conjunction with the ignition key if desired, such as in a fast re-start situation should the engine stall in traffic and it not be practical to remove the hands from the bars. Using either starting technique, however, it is advisable to disengage the clutch upon activating the starter since there is no lock-out on the gearbox and it is possible to start the engine in gear, with or without the clutch engaged. Should the clutch be engaged and the bike in gear, the motorcycle leaves—rider or no. Although the Magic Key concept is a masterful touch, brilliantly conceived, it has an Achilles' heel; the fragil aluminum casting which houses the tumbler lock and bolt. Since the Sport has only a centerstand—no sidestand—it was often parked leaning against a wall or handy tree. Just this pressure and a little movement were enough to fracture the bolt-on lock housing, necessitating its removal and welding. I recommend this lock housing be made of steel rather than aluminum in the interest of longevity. Not that the machine is difficult to raise up onto the centerstand; rather, it's very easy to do—unless you are recuperating from a broken back. One consolation; should a thief, realizing the above, attempt to snap the bars and break the lock, the lock remains immobilized more than ever, foiling his plan completely.

Except for the light and horn switches, all hand and foot controls are ideally positioned. But the turn indicator, dimmer switch and horn button are ergonomic nightmares. Ergonomics is the science of living things, the relationship between human beings and machines, and it is obvious that the engineer who designed these switches slept through the class. Additionally, he wired the headlight off-on switch independent of the ignition lock, so that any curious kid can come along when you're in watching a movie and leave your lights ablaze. And since the Sport relies strictly on electric starting, the owner could wind up doing the run and bump. Fortunately the battery is a massive brute about the size of that found in a Harley Electra Glide—12-volt, 32 ampere hours—with so much electrical overkill you'd have to take an axle to destroy it! Clutch and brake hand levers are ideally positioned and contoured. The clutch has a short friction point that will sucker you into some jackrabbit starts until you get used to the feel; after that the crisp, no-nonsense action may come to be preferred

Below—*Sanitary after 700 miles. Pedal linkage is expensive aircraft ball-end type. Most engine and chassis bolts have socket heads, making it easy to work on. Only oil filter is a sump screen that is accessible only by removing wet sump pan; should have quicker outside access and more sophisticated filter. As is, oil changes are recommended every 2000 miles. Gearbox oil cavity is separate.*
Bottom—*Foot pedals look alike; shift is on right, up for low. V-7 is exciting, superbly styled, of finest quality, handling and ride.*

over softer conventional actions. It proved-out that way for me.

Does the Guzzi V-7 qualify as a touring mount? You bet it does. With 32 of its 508 pounds dedicated to fuel supply, the Guzzi's 5.2-gallon tank provides a normal touring range of over 200 miles. The test model got between 37 and 42.5 miles per gallon at cruising speeds, depending upon the slant of the road and eagerness of the operator. And while our test came off without a hitch, had a fuse blown during the night, replacement would have been as simple as during broad daylight. For when the saddle is raised, a spring-loaded automatic trouble light comes on to illuminate the battery and fuse block area between the frame rails. The simple included light switch is identical to that in the glove compartment or trunk of your car, and it makes one wonder why other manufacturers don't wise-up to this luxury still foreign to the motorcycle establishment. Guzzi's employment of the multi-purpose ignition switch and this automatic trouble light dramatically emphasizes that motorcycling is still in the dark ages in so many ways. Still further evidence of their modern thinking—refusal to be mired in the motorcycle rut—is seen in the trick pivoted rear fender tip that can be elevated vertically by backing-off two friction knobs, allowing the rear wheel to be pulled straight out the back should tire repair be necessary. The sensible novelty of this roadster seems endless, and by now you should begin to feel a little more compassion for our kookie introductory paragraph.

Yes, in this day and age of exotic whiz-bangs the Moto Guzzi V-7 Sport stands out like the North Star. No need to free the clutch on a cold morning or wait five minutes for the engine to warm up to the point where it stops sneezing. Engine smoothness, roadability and ride are matchless when considered as an entity. The chain maintenance and rear tire problems that plague the big beasties don't exist in the Guzzi, nor does even a hint of driveline snatch. The low, sleek appearance and riding attitude create a standard all their own. If you prefer lower gearing—I don't—it is easily attained with an optional ring gear from Guzzi's El Dorado series that utilizes 37 teeth rather than the Sport's 35. The only alteration I would attempt—and the factory would do well to follow suit—would be the fabrication of an air cleaner box that would be large enough to permit retention of the high-performance carburetor velocity stacks while still purging the air flow of impurities. The machine is so outstanding that it is almost a moot point.

As corny as it may sound, you may appreciate my sentiments, scribbled inside a match book cover at one of several roadside stops during our happy acquaintance: "More than a motorcycle...a work of art." While saying it all, the phrase still seemed a little gross for the gang I run with. Then, just as I was about to pull away from a light, a young dude with a sweetie entwined around him pulled up alongside in a pickup truck: "Right on, man!" he breathed, as the two of them hung precariously out the window. Dammit, why couldn't I have said that! •

MOTO GUZZI SPORT V-7

TEST BIKE

Serial number........31723
Engine serial........31723
Date of manufacture........10/72
Base price as tested........$2489.
Factory warranty........3 months or 4000 miles

ENGINE

Type........90° V-twin four-stroke pushrod OHV
Displacement........748.4cc 45.66 cubic inch
Bore x stroke........82.5x70mm
Claimed HP @ RPM.....72 @ 7000
Claimed torque @ RPM........N.A.
Compression ratio........9.8:1
Lubrication system........Wet sump
Carburetion.....Two 30mm Dell'Orto square slide
Fuel required........Ethyl
Air filter........Optional
Ignition system...Battery, point, coil
Electrical system........Alternator, rectifier, regulator
Battery........12-V 32 Ampere Hours
Starting........Electric only
Exhaust........Twin downswept "Silentium"

DRIVE TRAIN

Primary/ratio........Gear/1.235
Clutch........Dry, 2-plate
Transmission........5-speed, constant mesh
Shift........Right foot, up for low
Gear ratios........2.0, 1.388, 1.047, 0.869, 0.750
Final drive........Shaft with constant velocity U-joint
front/rear/ratio........4.375
Overall ratios........10.806, 7.499, 5.657, 4.695, 4.052

CHASSIS AND SUSPENSION

Frame........Tubular, dual cradle
Fork........Telescopic, hydraulic damping
Caster/trail........28°/3.2 in.
Turn radius........91.5"
Steering damper........Hydraulic
Rear shocks.....Koni, constant rate spring. 3-way spring adjustment. Shock adjustable also for load & wear
Brakes,
front.......8.6" drum, 4 leading shoe cable-operated
rear........8.6 drum, 2 leading shoe
Tires,
front........3.25x18 Michelin rib
rear........3.50x18 Michelin universal
Balance weights........Yes
Rim locks, front/rear........No

WEIGHTS AND CAPACITIES

Weight, wet, unladen........508 lbs.
Weight bias, front/rear........236/272
Allowable gross weight........N.A.
Wheel loading front/rear........NA
Fuel capacity........5.2 gallons
Engine oil........7.3 pints
Gear oil........1 ¾ pints
Rear drive........12 ounces

STANDARD EQUIPMENT

Controls,
left hand..clutch, headlight, dimmer & horn switches
right hand........Brake & turn signal switch
Speedometer........160 mph No trip mileage reset
Tachometer........8000 rpm 7300 redline
Indicators........High beam, amps, oil, neutral
Locks........Ignition, fork, toolbox
Passenger provisions........Hand rails & pegs
Tools........15 pieces
Stands........Center stand only, Side stand optional

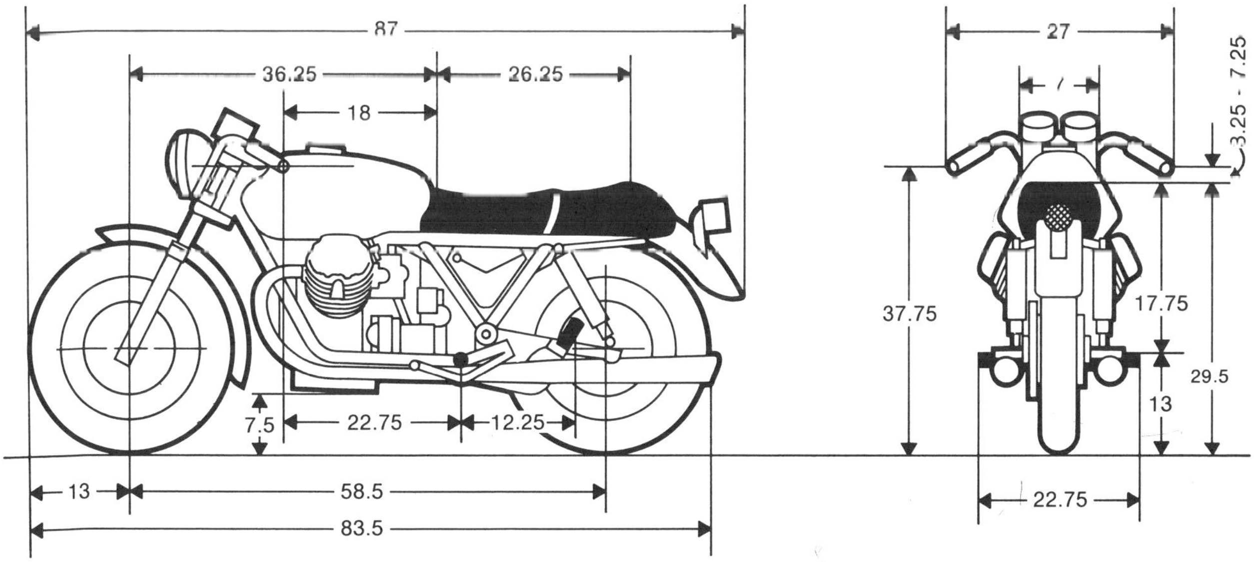

Continued from page 145

the lock home. Quite the best arrangement we have tried. The dualseat hinges to reveal an array of fuses with a courtesy light operated by plunger when the lights are on and the seat is lifted. A neat and thoughtful touch. In the headlamp are the usual array of lights one finds on Italian vehicles: yellow for neutral, red—oil, red—generator, green—lights. Although no flashing indicators are fitted to the Sports the wiring loom has provision for them and it would not take long to add them if one had a mind.

The general layout of the Moto-Guzzi vee-twin engine has had a good airing in these pages the past few years but we will recap, briefly, on the basic construction. It has a bore of 83mm. and stroke of 70mm. The Sports' compression ratio is 9.8 to 1 and the claimed maximum power 70 b.h.p., at 7,000 r.p.m. The crankshaft has bolted up connecting rods with plain bearings and is supported at either end by the usual hefty main bearings. At the front are the timing pinions, with the crankshaft pinion driving a half-speed camshaft pinion on top and the oil-pump pinion below. The oil-pump pinion drives a parallel gear type oil pump and is responsible for delivery by suction, with the return by gravity. Oil is filtered by a wire gauze unit in the bottom of the crankcase. In a separate housing in front of the timing gears a pulley on the extended crankshaft drives a belt to the Marelli generator mounted on top of the engine. Ignition timing can be checked, car style, by aligning marks on the generator pulley with a static mark on the timing case, At the opposite end of the engine the crankshaft flywheel is mounted in a separate housing. It is recessed to accept the two-plate clutch. Also at the rear, mounted on the right just behind the cylinder, is the distributor, driven by gear from the camshaft and with automatic advance and retard by bob-weights. The starter motor appears almost as something of an afterthought, mounted on the bottom left-hand side of the engine, beside the gearbox. It engages with a ring gear mounted between the flywheel and the gearbox. Carburettors are 30mm. Del 'Ortos. Much emphasis has been placed on the motor's "breathing" and two slightly untidy external flexible breather tubes run from the rocker housing. The final vent for the breather is located so that it points exactly at the rider's right ankle when he pulls up at traffic lights! That should not be difficult to overcome.

The five-speed gearbox provides overall ratios of 10.8, 7.59, 5.73, 4.78 and 4.09 to 1. Power is transferred by shaft with a double coupling. A rubber gaiter provides the necessary flexibility for the swinging fork, the righthand arm of the fork serving as a housing for the shaft. Final drive is by crown wheel and pinion (8:37).

Final statistics are: wheelbase, 57in; seat height, 30in; and ground clearance 11in. It needs a very well designed stand to enable anybody, short of Samson, to lift a 450 lb motorcycle effortlessly; and the V7's stand was only run of the mill, the missing lug making things no easier. No prop stand is provided, which is a pity.

There are those around who prefer the high-revving multi-cylinder brand of motorcycling to the Guzzi (and BMW) way of doing things. But others might think that a 750 c.c. vee-twin with shaft drive is the ultimate in motorcycling. And I should be inclined to agree with them. This machine is really one of the finest, most desirable, motorcycles on the market. The test machine returned a reasonable 48 m.p.g. overall but this figure would undoubtedly drop if one made full use of the considerable performance. It would be a very demanding rider, though, who found that the V7 could not satisfy his demands. My criticisms are few. Perhaps it would benefit from a disc brake on the front. The lights certainly could be better but there is nothing wrong with them that a Cibie or Marchal unit would not put right (though at £1,350 for the bike that should not be necessary).

The Moto-Guzzi enchanted me. It started easily, had the manners of a thoroughbred, the performance of a racer, handled well, stopped well, and was extremely comfortable. There are probably no more than half a dozen bikes worthy of the rather ridiculous title, "Superbike". The Moto-Guzzi V7 Sports, in my view, is one of them. A motorcycle of rare character, performance and quality.

B.P.